ASP Exam Prep 2026–2027

800 Realistic Practice Questions & Expert Rationales to Pass the BCSP Safety Professional Exam

Caldwell Abiodun

Author Introduction

Professional safety practice rests on the intersection of science, engineering, and human protection. As safety professionals prepare for certification, the challenge lies not only in memorizing standards but also in mastering the reasoning and applied judgment that underpins technical calculations and risk control. This guide has been designed to integrate those dimensions into a progressive learning framework that develops comprehension through analysis rather than rote recall.

Across nine domains, each section mirrors the emphasis of the official ASP11 blueprint. The narrative is technical yet approachable, reflecting the scientific principles that define contemporary safety practice—from mathematical analysis of hazards to regulatory interpretation and communication strategy. Domain 1 introduces the candidate to the analytical foundation of the field, where quantitative accuracy drives decision-making in complex work environments. Domains 2 through 9 expand this foundation into practical systems, environmental management, ergonomics, and ethics.

As the authoring framework for this study guide, prioritized structure, clarity, and proportional coverage. The study material draws exclusively on the verified BCSP outline and the user's provided reference sources, ensuring that every formula, explanation, and procedural description remains technically grounded. Each domain chapter follows a continuous instructional arc: from concept, to method, to applied example, and finally, to synthesis and review.

This edition also emphasizes reflective learning. Students are encouraged to apply each principle to real-world case scenarios—calculating, comparing, and interpreting outcomes as a professional would. When guided by this philosophy, the study process becomes an active rehearsal for certification-level competency.

Table of Contents

Copyright Page

The content is designed for individual study and instructional facilitation only. It does not substitute for BCSP official examination materials or any legal, regulatory, or procedural standard. Accuracy and completeness have been prioritized; however, learners are advised to cross-reference formulas, conversion factors, and procedural requirements with the official standards and guidance in their professional field.

Printed and distributed under the study development framework for 2025.

Examination Overview

The **Associate Safety Professional (ASP11)** examination represents a comprehensive evaluation of an individual's ability to apply scientific, mathematical, and technical reasoning to occupational safety practice. Administered by the **Board of Certified Safety Professionals (BCSP)**, the ASP credential forms a recognized benchmark for intermediate-level safety practitioners and serves as a prerequisite toward the Certified Safety Professional (CSP) designation.

The ASP examination blueprint defines the expected knowledge, skills, and abilities that support competent practice in safety engineering and management. The current ASP11 version, effective September 1, 2025, transitions fully to the 2025 professional standard and integrates the latest developments in risk assessment, process safety, and environmental controls. Each domain within the blueprint is weighted to reflect its proportional importance in daily professional performance. These weightings form the foundation for this study guide's page allocation and instructional emphasis.

Purpose of the ASP Certification

The ASP credential validates professional competence in identifying, evaluating, and controlling hazards that threaten people, property, and the environment. It emphasizes a balanced mastery of scientific analysis and practical implementation, equipping candidates to serve in roles requiring quantitative risk interpretation and system-level safety management. The examination ensures that certificants possess the technical fluency to make informed, ethical, and evidence-based decisions that protect life and maintain operational continuity.

Certification also affirms the candidate's commitment to continual professional development, adherence to ethical principles, and alignment with globally accepted standards for occupational safety and health.

Exam Administration and Format

The ASP11 examination is delivered in a **computer-based testing environment** at authorized centers. It consists of **200 multiple-choice questions**, including both scored and unscored items, administered over a **five-hour session**. All questions are designed to assess analytical application rather than rote memorization.

Each question presents a situational prompt requiring calculation, interpretation, or conceptual judgment based on the principles outlined in the nine domains. Some questions may require intermediate computation, graphical estimation, or the selection of the most appropriate formula from several possible approaches.

The candidate's final score is scaled to ensure consistent comparability across examination forms. Unscored items are indistinguishable from scored ones, reinforcing the need for thoroughness on every question.

Competency Focus

The ASP11 exam measures competency at three interrelated cognitive levels:

1. **Recall and Recognition** – understanding definitions, units, and basic laws governing physical and chemical behavior.

2. **Application and Computation** – performing calculations involving physics, mathematics, and exposure assessment using formulas and quantitative methods.
3. **Analysis and Interpretation** – integrating multiple data sources, determining causes, estimating risk magnitude, and making decisions that align with engineering controls and organizational policy.

To perform successfully, candidates must integrate quantitative reasoning with practical context—linking technical computation to safety outcomes. This ability to interpret data through professional judgment distinguishes the ASP-level practitioner from the entry-level safety technician.

Domain Structure and Weighting

The ASP11 blueprint defines **nine distinct domains**, each representing a core area of professional practice:

1. **Mathematical Calculations** – 10%
2. **Safety Programs and Concepts** – 25%
3. **Ergonomics** – 8%
4. **Fire Prevention and Protection** – 12%
5. **Emergency Preparedness and Response** – 10%
6. **Industrial Hygiene and Occupational Health** – 12%
7. **Environmental Management** – 7%
8. **Training, Education, and Communication** – 11%
9. **Legal** – 5%

This structure emphasizes both breadth and depth of understanding. Domains such as **Safety Programs and Concepts** carry greater weight because they encompass managerial integration and system application, while **Mathematical Calculations** and **Industrial Hygiene** form the analytical basis for measurement and evaluation. The balanced coverage ensures candidates demonstrate holistic competency—quantitative precision, managerial insight, and regulatory awareness.

Testing Conditions and Environment

Candidates are provided with an on-screen calculator and standard reference sheets during the exam. Familiarity with scientific notation, unit conversion, and formula manipulation is essential. Many items present practical scenarios such as dose calculations, ventilation system design, or control evaluation.

Environmental control of time is critical. The five-hour limit requires disciplined pacing—approximately **1.5 minutes per question**. Strategic time management includes recognizing which problems require computation and which are conceptually direct. Reviewing questions flagged for uncertainty near the end of the session is recommended once all items have initial responses.

The test environment maintains strict proctoring standards. Personal materials, communication devices, and unauthorized references are prohibited. Candidates should practice under simulation conditions to develop comfort with on-screen formula use and variable isolation without external aids.

Scoring and Performance Interpretation

Examination results are reported as **Pass/Fail** based on a criterion-referenced standard established through psychometric analysis. This ensures fairness across testing populations and consistency across versions. The scaled passing score corresponds to a statistically validated competency threshold rather than a fixed percentage.

BCSP periodically reviews and adjusts the passing standard to reflect evolving professional expectations. Thus, while practice test scores may serve as relative indicators, mastery should be measured by the ability to apply principles under unfamiliar conditions rather than by raw percentage alone.

Feedback provided to unsuccessful candidates identifies domain performance strengths and weaknesses, guiding further preparation. Candidates who do not pass may reapply according to BCSP's established policy framework.

Learning Objectives and Candidate Proficiency

Each domain of the ASP11 exam aligns with a defined set of learning objectives. Candidates are expected to:

- Demonstrate comprehension of technical terminology and formula components.
- Apply quantitative and qualitative analysis to problem-solving.
- Integrate knowledge of systems management, human factors, and legal requirements.
- Communicate reasoning clearly and logically when interpreting scenarios.

In practical terms, this means that mathematical and scientific understanding must support applied decision-making. The most successful examinees are those who transition fluidly between computation and professional judgment, reflecting not only theoretical mastery but real-world utility.

Preparation Strategy

Effective ASP preparation requires systematic engagement across domains rather than selective focus. Candidates should allocate study time according to domain weighting while recognizing interdependencies between subjects. For example, the mathematical principles in Domain 1 directly influence comprehension of Industrial Hygiene (Domain 6) and Fire Protection (Domain 4). Similarly, understanding system management in Domain 2 enhances readiness for questions in Training, Environmental Management, and Legal domains.

A balanced plan combines **conceptual study**, **problem-solving practice**, and **exam simulation**. Recommended steps include:

1. Reviewing foundational equations, units, and constants used throughout safety practice.
2. Performing sample calculations under timed conditions.
3. Reading procedural standards and linking them to quantitative outcomes.
4. Periodically summarizing learning in one's own words to strengthen memory retention.

Consistency of effort, not intensity alone, determines success. The ideal preparation period spans several months, with incremental review cycles that reinforce earlier material while integrating new domains.

Professional Scope and Responsibility

The ASP credential embodies accountability. Certified professionals are expected to uphold principles of integrity, evidence-based reasoning, and respect for human welfare. In the workplace, they act as translators of technical complexity into practical safety guidance—bridging science, operations, and management.

The examination evaluates readiness for that role. Each question tests not only factual accuracy but also ethical and contextual understanding. Candidates must interpret data without bias, prioritize prevention over convenience, and recognize that safety practice extends beyond compliance to continuous improvement.

Earning the ASP designation signifies preparedness for leadership grounded in competence. It marks the transition from knowledge accumulation to professional contribution—a commitment to safeguard others through informed analysis and responsible decision-making.

Integration into Career Progression

Achieving the ASP is not the culmination of study but a milestone in lifelong professional growth. The certification positions candidates to pursue advanced credentials such as the **Certified Safety Professional (CSP)** and to assume higher levels of operational authority. Employers regard the ASP as evidence of analytical ability, discipline, and ethical conduct—qualities that directly translate into workplace value.

Through the structured preparation embodied in this guide, candidates not only prepare for examination success but also strengthen their long-term professional framework. Each domain provides conceptual tools and practical reasoning that remain applicable throughout a career dedicated to protecting people, systems, and the environment.

Chapter 1 – Mathematical Calculations

Fundamentals of Measurement and Units

The precision of safety engineering depends on accurate measurement, and the foundation of all measurement lies in the consistent application of units. Every calculation used in industrial hygiene, environmental evaluation, or process control begins with the assumption that quantities are expressed in compatible systems. The International System of Units provides the standard reference, defining base dimensions such as mass, length, time, temperature, and amount of substance. Derived units are obtained from these bases through algebraic relationships that reflect the laws of physics. A lapse in unit integrity—whether in converting atmospheric pressure to Pascals or flow rate from cubic feet to cubic meters—can alter a result by orders of magnitude. Safety practitioners must not only convert correctly but also recognize the significance of dimensional homogeneity, where both sides of an equation maintain identical dimensions. This principle serves as a self-check mechanism for all calculations. In applied practice, mixed-unit systems remain common, particularly where legacy instruments use inches of water or cubic feet per minute. Careful dimensional analysis reconciles these older measures with modern standards. The reliability of exposure estimates, structural assessments, and process designs rests on this disciplined attention to units, ensuring that every computed value carries a consistent and interpretable meaning.

Scientific Notation and Significant Figures

Numerical representation in safety calculations requires balance between precision and practicality. Scientific notation allows large or small quantities—such as particle concentrations or reaction energies—to be expressed concisely without ambiguity. By representing numbers as products of coefficients and powers of ten, engineers avoid errors caused by misplaced decimal points and facilitate straightforward manipulation during multiplication or division. Precision, however, is not synonymous with accuracy. Significant figures indicate the certainty of measurement, conveying how much of a number's representation reflects actual knowledge rather than estimation. When performing sequential calculations, the rule is to carry all available digits through intermediate steps and round only the final result to the proper number of significant figures. This prevents propagation of rounding error and preserves the integrity of derived quantities. For instance, when calculating air sampling flow or the velocity of exhaust systems, rounding prematurely can misrepresent concentration levels or design thresholds. Understanding uncertainty through significant figures transforms raw numbers into credible information. In reporting results, the chosen precision should mirror the measuring instrument's capability—no more, no less. This disciplined numerical awareness is central to communicating technical results that can withstand scrutiny under regulatory and professional review.

Algebraic Relationships in Safety Calculations

Most safety-related computations derive from algebraic expressions linking physical quantities such as force, pressure, flow rate, and concentration. The ability to manipulate formulas is essential because problems often require isolating a variable from within complex relationships. Dimensional reasoning and symbolic substitution allow practitioners to adapt general equations to specific scenarios, such as calculating ventilation requirements or determining the dose of a contaminant over time. Linear and quadratic equations arise frequently in process control and exposure estimation, while exponential and logarithmic forms describe decay rates and contaminant dispersion. Algebra provides the framework for interpreting how changes in one parameter influence another, enabling predictive evaluation of system behavior. It also supports proportional scaling—estimating outcomes when conditions change while maintaining geometric or dynamic similarity. Mastery of algebraic manipulation ensures not only computational correctness but also conceptual understanding of the relationships that govern occupational safety phenomena. When integrated with measurement discipline and sound unit management, algebra becomes the language through which safety professionals translate abstract formulas into actionable design and evaluation decisions.

Dimensional Analysis and Conversion Techniques

Dimensional analysis is more than a mathematical convenience; it is the quality assurance tool of every calculation in the safety sciences. By expressing quantities in terms of their fundamental dimensions and tracing unit transitions through a problem, professionals can verify the logical structure of an equation before computing. Conversions between English and metric systems, between mass and weight, or between standard and actual gas conditions, all depend on recognizing how dimensions interact. A single calculation may involve pressure in atmospheres, volume in cubic feet, and flow in liters per second; without systematic conversion, the outcome would be meaningless. Factor–label methodology, where conversion factors are treated as fractions equal to one, provides a transparent path to the correct units. Each step in the analysis eliminates nonessential dimensions until the desired unit remains. This approach prevents errors, allows rapid verification, and strengthens confidence in derived results. In process safety, such verification safeguards against costly design failures. In industrial hygiene, it ensures that measured concentrations correspond to occupational exposure limits expressed in matching units. Dimensional analysis thus serves as both a diagnostic and preventive measure within the computational discipline of safety practice.

Graphical Representation and Data Visualization

The interpretation of quantitative data is often enhanced through visual representation. Graphs transform numerical relationships into perceptible patterns, allowing safety professionals to identify trends, outliers, and correlations that may remain hidden within tables. Line plots, scatter diagrams, and control charts convert statistical summaries into operational insight. A well-constructed graph respects proportional scaling, labeled axes, and accurate unit presentation, providing clarity without distortion. In exposure assessment, time–concentration curves illustrate patterns of variability and potential exceedance of limits. In quality control, run charts expose deviations that signal process instability. Graphical analysis also bridges the gap between raw data and decision-making by conveying complex findings to audiences with varied technical backgrounds. Modern computational tools facilitate rapid visualization, yet the responsibility for accuracy remains human. Every graphical element—whether a fitted regression line or a histogram interval—should reflect underlying data truthfully. Visual literacy, therefore, becomes a technical skill parallel to numerical computation. Through disciplined presentation, graphs support transparency and credibility, enabling data-driven safety management rooted in objective evidence.

Probability and the Nature of Uncertainty

In the realm of occupational safety, uncertainty is an inherent condition rather than an exception. Probability provides the framework for quantifying uncertainty and predicting the likelihood of events within defined boundaries. It bridges the gap between deterministic science and the variable realities of human systems and environmental conditions. By treating outcomes as random variables, probability theory enables practitioners to describe not only what may happen but how likely it is to occur. The concept of expected value translates probabilistic understanding into decision-making, allowing for rational prioritization of risk control measures. In exposure sampling, probability governs the interpretation of results—distinguishing between random fluctuation and true pattern. In process reliability, it defines the likelihood of failure over time. The mathematical rigor of probability empowers safety professionals to replace intuition with measurable reasoning, transforming qualitative perceptions of danger into quantitative assessments of risk. Recognizing that absolute certainty is unattainable, probability instead provides confidence, enabling action under uncertainty with calculated assurance.

Statistical Distributions in Measurement Analysis

Every measurement collected in safety and health practice is part of a distribution that reflects natural and procedural variability. Understanding the shape, spread, and central tendency of these distributions is critical for interpreting whether observed differences represent meaningful effects or random noise. The normal distribution, with its characteristic bell curve, underpins many inferential techniques because numerous independent influences combine to produce approximately normal outcomes. Yet other forms—binomial, Poisson, lognormal—describe phenomena such as count data, rare event frequency, and exposure variability. Selecting the correct distribution allows precise calculation of probabilities, confidence intervals, and control limits. In industrial hygiene, airborne concentration data often follow a lognormal distribution, requiring logarithmic transformation before statistical evaluation. Recognizing this prevents underestimation of extreme values and informs appropriate control measures. Statistical distributions transform individual measurements into population insight, linking observed evidence to broader inference. In safety engineering, this understanding forms the mathematical bridge between measured data and the decisions that protect human life.

Error Propagation and Measurement Uncertainty

All measurements carry error, and when those measurements are used in formulas, their uncertainties propagate through the calculation. The discipline of error analysis quantifies how imprecision in input values affects the confidence of results. By applying partial derivatives to express how sensitive a function is to each variable, one can compute the combined uncertainty in a derived quantity. This approach ensures that results are not interpreted with false precision. In safety practice, such analysis clarifies the reliability of calculated exposure limits, pressure drops, or chemical concentrations. Understanding propagation of error also informs instrument selection—revealing whether the measurement accuracy of one device complements or undermines the accuracy of another in a combined calculation. Uncertainty analysis promotes transparency, enabling professionals to communicate the range within which the true value likely lies. Rather than a limitation, it serves as a safeguard, preventing overconfidence and ensuring decisions rest on honest interpretation of data. The mastery of error propagation marks a transition from calculation to professional judgment.

Application of Statistics in Safety Evaluation

Statistical reasoning transforms isolated measurements into meaningful evaluation of safety performance. By summarizing data through measures of central tendency and dispersion, practitioners can identify typical values and understand variability. Statistical inference extends this capability, enabling conclusions about entire populations from limited samples. Confidence intervals provide ranges within which true parameters are likely to fall, while hypothesis testing distinguishes between apparent and genuine differences. In industrial hygiene, these tools determine whether observed exposure levels exceed regulatory limits with statistical significance. In process control, they identify deviations that may indicate emerging hazards. Proper use of statistics requires both numerical skill and interpretive insight—the ability to discern what the numbers imply about real-world conditions. Misapplied statistics can obscure truth as easily as reveal it; therefore, the safety professional must pair technical calculation with ethical communication, ensuring that statistical results inform action rather than justify complacency. Through sound statistical application, data become evidence, and evidence becomes safety improvement.

Mathematical Modeling in Safety Systems

Mathematical models provide simplified representations of complex systems, allowing prediction, optimization, and evaluation of safety performance. Whether modeling gas dispersion, thermal stress, or chemical reaction kinetics, equations serve as conceptual laboratories where conditions can be tested without physical risk. Models combine theoretical laws with empirical data, translating observed behavior into quantitative relationships. Their strength lies in the ability to forecast outcomes under hypothetical scenarios—such as the effect of ventilation changes on contaminant distribution or the expected injury rate following procedural modifications. However, models are only as valid as their assumptions. Simplifications regarding uniform mixing, steady-state flow, or constant temperature must be evaluated against real conditions. Calibration and validation using measured data ensure the model remains grounded in reality. Within safety management, modeling supports proactive control, allowing decision-makers to anticipate rather than react. It converts abstract principles into operational foresight, exemplifying the analytical maturity expected of an ASP-certified professional.

Integration and Application Summary

Mathematical calculations form the structural foundation of safety science. From the conversion of units to the modeling of system behavior, they transform physical phenomena into quantifiable understanding. The discipline requires more than computation—it demands precision of language, awareness of uncertainty, and respect for dimensional consistency. In professional practice, each equation links directly to a human outcome, whether controlling exposure, predicting equipment failure, or ensuring regulatory compliance. Mastery of mathematics in this context is mastery of clarity itself: the ability to describe, predict, and justify decisions that protect people and processes. As the first chapter of this study guide, Mathematical Calculations establishes the analytical discipline that supports every domain that follows, ensuring that the science of safety remains both accurate in number and honorable in purpose.

Chapter 2: Safety Programs and Concepts

Evolution of Safety Program Philosophy

Modern safety management evolved from rule-based compliance to risk-based prevention. Early programs focused primarily on injury statistics and enforcement, but over time, industries recognized that compliance alone could not sustain safety performance. The emergence of systemic frameworks, such as Safety Management Systems (SMS), integrated human factors, engineering design, and organizational culture into proactive hazard control. According to the provided material, this evolution paralleled the growth of management systems in quality and environment, establishing safety as a core organizational process rather than a peripheral function. The shift toward integrated systems—exemplified by the Safety and Health Management System (SHMS) approach—brought structure to leadership responsibility, hazard identification, and continuous improvement. By redefining safety from an obligation into a strategic management priority, organizations now use data-driven performance measures, internal auditing, and risk evaluation to anticipate failures before they result in harm. The modern safety program thus functions as an adaptive ecosystem, balancing compliance with innovation and embedding safety within every operational decision.

Foundations of a Safety Management System

A Safety Management System (SMS) provides a structured method for managing safety risk, ensuring that operational and managerial processes systematically reduce potential harm. The framework typically includes four essential components: policy, risk management, safety assurance, and safety promotion. Within these, management commitment establishes accountability, while systematic hazard analysis forms the core of risk control. The materials describe that a robust SMS integrates compliance verification with performance validation—ensuring legal adherence while assessing practical effectiveness. This framework aligns with international standards that require leadership engagement, communication, and continuous feedback loops. A distinguishing feature is its adaptability; regardless of industry, the same structure—policy, organization, planning, implementation, and review—guides consistent improvement. The SMS concept reflects a balance between reactive and proactive functions: it not only investigates incidents but predicts and prevents their recurrence through data trend analysis and cultural reinforcement. In mature programs, the SMS becomes a living system that evolves with organizational learning, merging operational intelligence with human reliability management.

Management Leadership and Accountability

Effective safety programs depend upon visible and sustained management leadership. Leadership defines the tone of organizational safety culture through policy declarations, strategic objectives, and tangible resource allocation. In the provided reference materials, management leadership is identified as the cornerstone of every safety and health management system, where accountability extends from senior executives to supervisory levels. The demonstration of leadership occurs through active participation in safety planning, performance evaluation, and communication of expectations. By establishing measurable goals and integrating them into corporate scorecards, management ensures that safety performance carries equal weight with productivity. Furthermore, leaders are expected to act as role models, adhering strictly to safety protocols and encouraging employee engagement. A critical insight from the system manuals emphasizes that leadership without accountability degenerates into symbolism; thus, authority must be matched with measurable responsibility. When leaders incorporate safety into budgetary and operational decisions, the organization transitions from compliance-driven behavior to genuine cultural commitment, sustaining long-term performance improvements.

Employee Involvement and Organizational Culture

Employee participation transforms safety from a top-down directive into a shared responsibility. The referenced safety management system handbooks consistently stress that employees hold unique situational knowledge that management alone cannot access. Structured mechanisms such as joint safety committees, observation programs, and open reporting channels encourage participation and build trust. Organizational culture—the shared beliefs about safety—evolves through this collaboration. According to the hospital SHMS framework, engagement empowers staff to identify hazards early, propose corrective actions, and validate solutions in practice. This participative approach reduces resistance to change and embeds ownership across departments. The human factors perspective also suggests that culture influences perception of risk; therefore, communication and recognition programs are essential to reinforcing desired behaviors. A positive culture integrates error management with learning, focusing on improvement rather than blame. Through continuous dialogue and empowerment, employees evolve from passive subjects of policy into proactive contributors to organizational safety maturity.

Hazard Identification and Risk Assessment

Hazard identification is the analytical backbone of all safety programs. It involves recognizing conditions, processes, or behaviors that could cause injury, illness, or damage, followed by systematic evaluation of their likelihood and consequence. The ICAO framework and OSHA guidance both define this as a continual, iterative process. Assessment techniques range from job safety analyses and failure mode evaluation to probabilistic risk modeling. The aim is not to eliminate all hazards but to maintain risks at levels that are as low as reasonably practicable. In comprehensive programs, risk registers document findings, and control measures are prioritized based on severity and frequency. Moreover, modern risk assessment includes psychosocial and ergonomic factors, acknowledging that cognitive overload and fatigue can equal physical dangers. Integrating quantitative methods with observational data allows the organization to move from intuition-based judgments to evidence-driven control strategies. The process becomes cyclical: identify, evaluate, control, and revalidate—forming the safety program's continuous improvement loop.

Hazard Control and Preventive Engineering

Once risks are assessed, preventive engineering transforms analysis into action. The hierarchy of controls—elimination, substitution, engineering control, administrative measures, and personal protection—provides a structured pathway for intervention. The industrial safety management materials illustrate that successful control depends on design foresight as much as operational discipline. Engineering solutions, such as machine guarding, ventilation, or automation, eliminate exposure at its source, representing the most reliable control level. Administrative strategies, including rotation schedules and permit-to-work systems, address residual risks. Preventive maintenance, inspection, and change management ensure that safety design persists through operational aging. The integration of hazard prevention with production engineering creates dual accountability: designers must anticipate misuse, and operators must preserve control integrity. A key insight is that hazard control should not be reactive; it must be built into project planning, procurement, and process design from the outset. In this sense, engineering prevention embodies the philosophy of safety by design.

Training, Competence, and Human Reliability

Competence development anchors the safety system in human capability. Every program relies on well-trained personnel who understand both procedures and underlying risk principles. The Safety Management System guides specify education as a continuous function—not a one-time orientation. Effective training aligns with job-specific hazards, regulatory obligations, and the organization's strategic objectives. Competence is verified through assessments and reinforced by on-the-job supervision. The training process also supports behavioral reliability: by developing situational awareness, decision-making skills, and procedural discipline, workers become more resilient to operational stress. The provided handbooks emphasize dual-level education—technical instruction for execution and managerial training for oversight. Modern systems integrate simulation, case-based learning, and digital tracking to measure knowledge retention. When combined with periodic evaluation, this creates a feedback-driven competence loop. The ultimate measure of training effectiveness is demonstrated not merely in test results but in consistent safe behavior during routine and non-routine operations.

System Evaluation and Continuous Improvement

Continuous improvement transforms a static safety program into an adaptive system. Evaluation mechanisms, including audits, inspections, and management reviews, ensure that implemented controls remain effective under changing conditions. The VTT publication on safety management auditing notes that reliability in evaluation depends heavily on auditor competence and structured criteria. Internal audits verify compliance, while external audits validate system effectiveness and objectivity. Findings generate corrective and preventive actions, which are then tracked through performance indicators. Feedback loops from incident investigations and employee reports feed into the same improvement cycle. Importantly, improvement extends beyond rectifying deficiencies—it involves innovation, benchmarking, and process redesign. The system learns by comparing current performance to leading practices and regulatory expectations. In this context, auditing evolves from a policing function into an organizational learning tool, strengthening both compliance and cultural maturity through disciplined reflection and transparent performance review.

Incident Investigation and Corrective Actions

Despite preventive efforts, incidents may still occur, and their investigation represents a pivotal opportunity for learning. Effective systems treat incidents as data-rich events that reveal underlying system weaknesses rather than isolated human errors. According to the referenced safety manuals, structured methods such as root cause analysis, event tree mapping, or the Reason model are employed to trace organizational contributors. The aim is not to assign blame but to identify failures in design, supervision, or communication. Corrective actions must address both immediate causes and systemic contributors, ensuring recurrence prevention. Timely communication of lessons learned reinforces transparency and strengthens cultural resilience. Integration of findings into training, policy updates, and engineering review closes the loop between learning and improvement. A well-managed investigation process builds trust, as employees recognize that reporting leads to constructive change. Thus, incident investigation becomes a cornerstone of organizational intelligence within the broader safety management framework.

Integration, Auditing, and Strategic Safety Performance

The final dimension of an effective safety program is its integration with overall business strategy. Safety management cannot function in isolation; it must align with quality, environment, and operational excellence initiatives. Integration ensures shared data systems, unified performance metrics, and harmonized improvement processes. Auditing provides the mechanism for this integration by translating safety outcomes into measurable organizational performance. The technical research materials emphasize that auditing reliability and validity determine whether management decisions reflect true system status. Strategic performance monitoring employs indicators such as total recordable incident rates, near-miss trends, and control verification outcomes. At the executive level, safety performance becomes a key determinant of sustainability and corporate responsibility. When safety metrics are linked to planning and budgeting, they reinforce a culture of accountability. Ultimately, the integrated safety program transcends compliance to become a strategic asset—driving operational reliability, workforce morale, and public trust.

Chapter 3 – Risk Assessment and Control

Principles of Risk Management in Safety Practice

Risk management is the structured process by which organizations identify, evaluate, and control potential events that could cause injury or loss. The fundamental objective is to establish a systematic approach that anticipates harm before it occurs. Within the safety profession, risk management combines analytical methods with managerial decision-making, forming the bridge between technical evaluation and operational control. A risk is defined by the combination of probability and consequence; therefore, both must be quantified or qualitatively assessed. The modern framework aligns with the hierarchy of control, recognizing that elimination or substitution of hazards is superior to administrative or personal protective measures. Effective programs integrate assessment outcomes into planning, budgeting, and continuous improvement. Risk management is cyclical: assessment informs action, and feedback from control performance refines future evaluations. It also demands communication—clear articulation of risk levels ensures that decisions are transparent and accountable. As safety programs mature, the focus shifts from isolated hazard control toward total risk governance, embedding prevention within the core of every operational process.

Hazard Recognition and Classification

Hazard recognition precedes all control efforts. It requires a disciplined evaluation of the workplace environment, processes, and human interactions to reveal conditions that may result in harm. Hazards are typically grouped into physical, chemical, biological, ergonomic, and psychosocial categories, each requiring distinct detection methods. Visual inspections, job hazard analyses, and process mapping remain fundamental techniques. However, true hazard recognition extends beyond observation—it incorporates data from incident reports, exposure monitoring, and employee feedback. In industrial settings, OSHA's technical guidance provides frameworks for recognizing complex exposures such as polymer composites, solvents, and reactive agents. Ergonomic hazards demand attention to task repetition, posture, and force application, drawing on anthropometric data to match human capability to job design. Classification enables prioritization; high-severity, high-likelihood hazards receive immediate attention, while less critical risks are managed through scheduled review. Recognizing hazards as dynamic rather than static phenomena ensures that changes in process, equipment, or workforce composition trigger re-evaluation. This vigilance transforms hazard recognition from a periodic inspection task into a continuous organizational reflex.

Qualitative and Quantitative Risk Assessment Methods

Risk assessment methods fall broadly into qualitative and quantitative approaches. Qualitative assessments use descriptive scales—low, medium, high—to estimate likelihood and severity when data are limited. They are useful for rapid decision-making and prioritization but require expert judgment to remain consistent. Quantitative methods, in contrast, apply numerical values and statistical models to predict frequencies and outcomes. Techniques such as fault tree analysis, event tree analysis, and probabilistic risk assessment quantify complex failure pathways. According to OSHA and NIOSH methodologies, integrating both approaches yields balanced insight—qualitative for broad screening and quantitative for detailed evaluation. Quantification also supports cost–benefit analysis by assigning measurable impact to potential losses. Regardless of method, transparency in assumptions and data sources remains essential. The chosen approach must suit the hazard's complexity and the decision's significance. Properly applied, assessment methods provide a defensible basis for selecting controls and verifying effectiveness, ensuring that safety actions rest on evidence rather than intuition.

Ergonomic Risk Assessment and Human Factors Integration

Ergonomic assessment extends traditional risk analysis to the interface between human capability and task demand. Using insights from *Fitting the Human* and NIOSH's ergonomic guidelines, practitioners analyze posture, force, repetition, and environmental conditions that contribute to musculoskeletal disorders. Manual material handling, workstation design, and task variability are measured through observation, anthropometric comparison, and, where possible, biomechanical modeling. Ergonomics interprets physical data—such as lifting indices or reach envelopes—into practical design recommendations that minimize fatigue and overexertion. Risk control in this field favors engineering redesign over personal behavior modification. For example, repositioning materials to waist height or introducing mechanical assists yields greater reduction in strain than training alone. Human factors engineering broadens the scope to include cognitive workload, perception, and decision-making under stress. Together, these disciplines transform risk assessment into a human-centered process, ensuring that systems accommodate people rather than expecting people to adapt to poorly designed systems.

Exposure Assessment and Environmental Monitoring

Exposure assessment quantifies how much of a hazard a worker encounters over time. It applies particularly to chemical and physical agents such as dusts, vapors, noise, and radiation. The OSHA Technical Manual outlines standard methods—personal sampling, area monitoring, and biological indices—to estimate exposure intensity and duration. The goal is to compare measured values to occupational exposure limits and evaluate compliance. Continuous monitoring technologies enhance this process by detecting fluctuations that short-term sampling might miss. Environmental monitoring extends assessment to ventilation efficiency, thermal comfort, and illumination—all of which influence physiological and psychological performance. In risk control, data from exposure assessments guide ventilation design, substitution of less hazardous materials, and selection of personal protective equipment. When integrated with health surveillance, exposure monitoring provides early warning of systemic risk, allowing intervention before symptoms arise. Thus, exposure assessment functions as both diagnostic and preventive science within comprehensive safety management.

Risk Control Strategies and the Hierarchy of Control

Effective control strategies stem from a clear hierarchy: elimination, substitution, engineering control, administrative control, and personal protective equipment. This structure prioritizes inherent safety—designing out hazards before relying on human behavior. For instance, replacing a solvent-based resin with a water-based alternative eliminates a class of exposure rather than mitigating it. Engineering controls such as enclosures, interlocks, or automation isolate people from hazards. Administrative measures—procedures, rotation schedules, and work permits—manage exposure where elimination is impractical. PPE serves as the last barrier, essential but inherently limited by fit and compliance. Each level of control interacts; true effectiveness depends on integration rather than reliance on a single measure. The ergonomic principle of "designing for the human" complements this hierarchy by ensuring controls do not introduce new strains or cognitive burdens. Risk control, therefore, is not static compliance—it is the continual refinement of system design to reduce vulnerability without compromising operational performance.

Human Reliability and Behavioral Safety

Even the most sophisticated controls depend on human reliability. Human reliability analysis evaluates the probability that individuals will correctly perform required actions within defined conditions. It recognizes that error arises not from negligence but from systemic factors—complex interfaces, ambiguous procedures, or excessive workload. Behavioral safety programs seek to strengthen reliability through observation, feedback, and reinforcement of safe acts. The ergonomics literature stresses that sustainable behavior change emerges from supportive environments rather than punitive monitoring. Training, participatory design, and empowerment contribute to reducing human error probability. Advanced analysis tools model task steps and identify critical points where performance deviation could lead to failure. Integrating these insights into risk assessment allows designers to build error tolerance into systems—using fail-safes, warnings, and simplified controls. By viewing humans as assets within the safety system rather than sources of failure, organizations enhance resilience and align individual behavior with systemic reliability.

Control Verification, Auditing, and Continuous Improvement

Verification confirms that implemented controls perform as intended. Regular audits, inspections, and performance reviews close the feedback loop between assessment and outcome. OSHA's evaluation models emphasize measurable criteria—exposure reduction, compliance rate, and incident frequency. Auditing compares these metrics against established benchmarks, revealing whether control measures remain effective or require adjustment. Continuous improvement depends on systematic review cycles that transform findings into new objectives. The Plan–Do–Check–Act methodology is particularly suited to safety programs, promoting iterative refinement. Feedback from ergonomic assessments, incident reports, and worker suggestions provides the qualitative dimension of verification, ensuring that technical success aligns with human usability. When verification becomes routine rather than reactive, risk control evolves into a self-correcting mechanism capable of adapting to organizational change, technological advancement, and emerging hazards.

Documentation, Communication, and Decision-Making

Accurate documentation anchors the entire risk assessment process. It records methodologies, assumptions, results, and decisions, forming both a legal defense and a learning archive. Communication ensures that this information reaches those who need it—operators, supervisors, and executives alike. Clear presentation of risk levels and control measures fosters informed decision-making and accountability. The process must avoid over-technical language; the aim is comprehension, not confusion. Visual matrices and simplified summaries often convey risk hierarchy more effectively than tables of numbers. Decision-making under uncertainty benefits from structured communication, where transparency about limitations and residual risk guides responsible judgment. In mature safety cultures, communication becomes two-way: management disseminates information, and workers provide feedback from practical experience. This exchange transforms documentation from a static record into a dynamic management tool, sustaining trust and collective ownership of safety outcomes.

Integration of Ergonomic Design into Risk Prevention

The final stage of risk control merges ergonomic design with strategic prevention. The goal is to create environments where risk is minimized by default—where tasks, tools, and processes inherently support safe behavior. The ergonomic frameworks from *Fitting the Human* and *Ergonomics in Action* advocate designing workplaces that fit diverse users, accommodate physical variability, and reduce unnecessary cognitive demand. By integrating ergonomics at the design stage, organizations prevent risks from emerging rather than attempting to manage them later. Practical examples include adjustable workstations, optimized material flow, and reduction of manual handling through automation. These measures simultaneously improve productivity and worker well-being. The integration process requires collaboration among engineers, safety professionals, and end-users, ensuring that solutions are functional and sustainable. Ultimately, ergonomically informed design represents the highest maturity of risk control—prevention embedded in structure, culture, and everyday operation.

Chapter 4 – Fire Prevention and Explosives Control

Fire science and the professional scope of protection

Understanding fire at a professional level begins with recognizing it as a rapid, self-sustaining chemical reaction that releases heat, light, and reaction products. For the safety practitioner, this scientific understanding informs practical control: identifying fuels, oxidizers, and ignition mechanisms that together enable combustion. Fire behavior depends on the characteristics of the fuel package, the geometry of the space, ventilation patterns, and transient phenomena such as flashover and backdraft. Industrial settings introduce complexity through chemical inventories, process heat, and enclosed spaces where fire growth can be very rapid. The professional role therefore spans prevention, detection, mitigation, and recovery, and requires harmonizing engineering controls, administrative strategies, and emergency preparedness. Fire science also underpins selection criteria for suppression systems and protective construction, because the required response—early detection, confinement, or full suppression—depends on expected heat release rates and smoke production. Importantly, modern fire protection is risk-based, integrating probabilistic assessment with prescriptive codes. The practitioner must translate code requirements into site-specific protection strategies that consider business continuity and life safety together. This integrated stance recognizes that fire protection is not an add-on but an essential element of process design and organizational resilience.

Fire chemistry, heat release and growth dynamics

At its core, a fire is a chain reaction in which fuel molecules oxidize and liberate energy. Heat release rate is the principal metric that determines how quickly a fire grows and how difficult it will be to control. Small differences in fuel characteristics and ventilation produce dramatically different growth curves: a smoldering ignition may persist for a long period, while a well-ventilated hydrocarbon pool fire can escalate to flashover in minutes. Heat transfer mechanisms—conduction, convection, and radiation—govern fire spread between materials and compartments. Radiative heat flux is especially important during growth toward flashover, because it can preheat remote fuel surfaces until they ignite. The notion of a fire timeline, from ignition through growth, possible flashover, and decay, guides protection strategy. Early detection and rapid suppression aim to interrupt the growth phase, preventing conditions where occupant egress or firefighting becomes untenable. Design decisions such as compartmentation, rated separation, and ventilation control alter the available oxygen and thermal pathways, and therefore change growth dynamics. Quantitative understanding—heat release rate per unit area, flame temperatures, smoke yield—translates directly into engineering choices for suppression capacity, sprinkler spacing, and structural fire resistance.

Characteristics of flammable liquids, gases, and vapors

Flammable liquids and gases present unique hazards because their physical state and vaporization behavior determine how they mix with air and ignite. Key properties such as flash point, boiling point, vapor density, lower and upper flammability limits, and vapor pressure together define the conditions under which a material poses a fire or explosion risk. Low flash point liquids easily form ignitable vapors at ambient temperatures and therefore require rigorous storage and handling controls. Gases with vapor densities greater than air may accumulate in low points, while lighter gases tend to collect under ceilings or within enclosed structures. Vapor cloud dispersion is influenced by source geometry, release rate, and atmospheric conditions; confined vapor clouds present acute explosion hazards if ignitable mixtures develop. Practical controls stem from acknowledging these properties: ensuring adequate ventilation, using inherently safe temperatures and pressures where possible, and implementing ignition source control. Selection of compatible piping, bonding and grounding for liquids with low flash points, and gas detection systems further mitigate the likelihood that a flammable atmosphere will form or be ignited.

Combustible dusts and explosive atmospheres: recognition and behavior

Combustible dusts create a ubiquitous but sometimes overlooked explosion hazard across many industries. When fine particles are suspended in air at sufficient concentrations, they form dust clouds that, given an ignition source and confinement, can result in violent explosions. The severity depends on particle size distribution, moisture content, composition, and cloud concentration. Primary dust deflagrations may dislodge accumulated dust layers and trigger secondary, often more destructive explosions. Recognizing dust hazard requires evaluating both process operations and housekeeping: material transfer points, sieving operations, and enclosed conveyors are common locations for dust cloud generation, while poorly controlled accumulations on ledges create secondary hazard potential. Practical mitigation emphasizes prevention through process enclosure and local capture, ventilation design that avoids cloud formation, and strict housekeeping to eliminate accumulations. Where exposure cannot be eliminated, explosion protection techniques such as venting, suppression, and isolation are applied according to standardized approaches. As with gases and liquids, the professional must interpret material testing data—Kst values, minimum ignition energy, and maximum explosion pressure—to align protective measures with the specific explosibility characteristics of the dust.

Ignition sources and strategies for their control

Ignition sources are diverse, ranging from open flames and hot surfaces to electrostatic discharge, frictional heating, mechanical sparks, and electrical faults. The control philosophy addresses the elimination, isolation, or management of these sources relative to potential fuel and oxygen availability. Engineered prevention begins with inherently safer design: locating potential ignition-producing equipment outside hazardous zones, selecting intrinsically safe electrical devices, and specifying appropriate equipment grounding and bonding for hydrocarbon handling. Administrative controls include hot work permits, maintenance isolation procedures, and hot surface monitoring. For processes where ignition cannot be fully eliminated, detection and rapid suppression systems provide the compensating protection. The professional must also consider latent ignition sources such as inadvertent energy stored in compressed systems or the thermal inertia of process equipment during shutdowns and restarts. Effective ignition control integrates design, procedure, and training, reducing both the likelihood of ignition and the chance that ignition will encounter a fuel-rich, confined atmosphere.

Fire prevention through process design and safe storage practices

Prevention begins at the design stage and extends through procurement, operations, and maintenance. Designing out hazards may mean substituting less flammable materials, minimizing onsite inventory of flammables, or altering process conditions to reduce vapor pressure and temperature. Storage practices follow prescriptive guidance that groups materials by hazard classification, limits maximum allowable quantities per control area, and prescribes construction and drainage features for containment. Mechanical integrity programs ensure tanks, piping, and valves remain sound, preventing leaks that could create hazardous atmospheres. For flammable liquids, proper classification and segregation, secondary containment, and spill control are critical. For pressurized gases, robust leak detection and relief systems are essential. Equally important is change management: design modifications, maintenance activities, and production scale-ups can inadvertently increase fire risk unless evaluated against the facility's fire protection strategy. Prevention therefore links engineering design with operational discipline, enforcing controls through permits, inspections, and documented procedures that keep the implemented protections aligned with the workplace reality.

Detection, alarm systems, and smoke management

Early detection and prompt notification are the keystones of life safety and initial fire control. Detection technologies—ionization and photoelectric smoke detectors, heat detectors, flame detectors, and gas sensors—have distinct strengths and limitations that dictate appropriate application. Smoke detection excels in early smoldering or flaming scenarios with rapid smoke production, while heat detectors are suited for harsh environments where smoke sensors might nuisance-alarm. Alarm systems must reliably transmit a clear signal to occupants and the fire service, and must be coordinated with automatic response features such as HVAC shutdown and fire suppression system actuation. In complex occupancies, smoke control systems manage smoke movement by pressurization, exhaust, and compartmentalization to preserve means of egress and firefighter access. The design of detection and smoke management must consider expected fire growth, ventilation patterns, and the probability of detector impairment. Routine testing, maintenance, and documented inspection schedules ensure detection systems remain effective over the life of the facility. When detection is integrated with building automation and the fire department interface, response times shorten and overall outcomes improve.

Fire suppression systems: selection and application

Suppression systems translate risk assessment into tangible capacity to mitigate a fire before it grows uncontrolled. Automatic sprinkler systems remain the most widely applicable means to contain and often extinguish fires in occupancies ranging from industrial warehouses to office spaces. System selection and density are guided by expected heat release and fuel loading, and choices include wet pipe, dry pipe, preaction, and deluge systems according to hazard and ambient conditions. For special hazards, such as electrically energized equipment or sensitive archives, gaseous suppression agents provide extinguishment without water damage, while foam systems are indispensable for hydrocarbon pool fires and hydrocarbon tank protection. Portable extinguishers complement built-in systems by enabling occupant response to incipient fires when personnel are trained and conditions are safe. Equally important are system redundancy, water supply reliability, and maintenance regimes that assure availability. Design must also consider interaction with detection and egress systems; for example, activation of a deluge system may require coordinated evacuation and isolation of electrical power. Selecting suppression thus balances fire dynamics, property considerations, and human factors to maximize protection without introducing new hazards.

Explosion protection: venting, suppression, isolation, and mitigation

Explosion protection is a multi-layered engineering discipline that addresses both primary and secondary explosion scenarios. For dusts and vapor clouds, strategies include venting to reduce pressure buildup, suppression systems that detect incipient combustion and quench flame propagation, and isolation methods such as fast-acting valves and rotary valves that prevent flame transmission between process units. Structural design can incorporate reinforced construction or explosion panels to direct overpressure away from occupied zones. For process vessels, properly sized relief and rupture systems prevent catastrophic vessel failure. Equally critical is preventing ignition of a flammable cloud in the first place through ventilation and concentration control. Secondary explosions are avoided by eliminating dust accumulations and isolating ignition-prone transfer points. The professional must rely on standardized explosion test data and design guidance to select appropriate protection measures and to document their expected performance. Because explosion consequences can be disproportionately large relative to their initiating event, conservative treatment and rigorous verification are prudent.

Emergency planning, egress, and integration with fire response

Even the best preventive systems may not avert every incident, so robust emergency planning and egress design are indispensable. Emergency plans translate technical protection measures into human action: alarm notification, evacuation routes, designated assembly areas, and communication protocols with emergency responders. Means of egress planning addresses occupant load, travel distance, stair and corridor capacity, and accessibility, ensuring that escape remains feasible under smoke and heat conditions. Fire brigades—internal or municipal—must be provided with information on hazardous inventories, access points, control valves, and utility shutoffs to streamline intervention. Training and drills build institutional memory and reduce the confusion inherent in emergencies. Post-incident actions include investigation, corrective action implementation, and learning loops that feed into process safety and facility design improvements. Integration of emergency plans into business continuity and recovery ensures that organizational resilience extends beyond life safety to operational recovery, preserving both human welfare and the viability of the enterprise.

Chapter 5 – Emergency Preparedness and Response

Foundations of Emergency Management and Preparedness

Emergency preparedness represents a deliberate, organized approach to reducing vulnerability and ensuring resilience against disruptive events. It extends beyond regulatory compliance to a holistic management process integrating people, infrastructure, and information. Preparedness begins with recognition that emergencies—whether natural, technological, or human-induced—are inevitable, but their consequences can be controlled. NFPA 1600 defines preparedness as a continuous cycle of planning, training, exercising, evaluating, and improving. The foundation rests on establishing policies that define leadership responsibility, resource allocation, and communication pathways. Planning addresses both immediate life safety and long-term continuity objectives. It also requires understanding the specific hazard landscape—geographic, operational, and social—that shapes exposure. Preparedness is not a static plan but a culture of readiness built through participation, practice, and reflection. Every organization, regardless of size, must ensure that emergency functions integrate into daily operations so that activation during crisis is seamless rather than improvised. This integration transforms preparedness from documentation into capability.

Hazard Identification, Risk Assessment, and Impact Analysis

The first practical step of emergency management is systematic hazard identification and risk assessment. This process, rooted in the same principles as occupational safety risk evaluation, expands the focus to include external and cascading threats. Hazards may arise from natural events such as floods, storms, or earthquakes; technological incidents such as chemical releases or utility failures; and human-caused crises such as violence or terrorism. The assessment quantifies both likelihood and consequence while examining vulnerability—how exposed assets, processes, and people are to these hazards. Impact analysis extends the evaluation to secondary effects such as supply chain disruption, data loss, and community interdependence. Tools such as hazard vulnerability analysis matrices and business impact assessments translate abstract risk into prioritized action. The outcome is a clear profile of critical functions and resources requiring protection. This baseline supports mitigation strategies and continuity planning, ensuring that limited resources are directed to areas where failure would have the most severe consequences.

Emergency Planning and Program Development

An effective emergency plan transforms analysis into operational structure. It defines roles, responsibilities, and procedures for preventing, preparing for, responding to, and recovering from incidents. NFPA 1600 establishes that such a plan must be comprehensive yet flexible, addressing the all-hazards spectrum and coordinating with community resources. Program development begins by defining policy statements endorsed by senior leadership, establishing authority for plan activation, and identifying succession for key roles. Planning extends beyond internal coordination to external stakeholders—contractors, emergency services, suppliers, and neighboring facilities—since crises rarely respect organizational boundaries. Plans must specify notification methods, communication hierarchies, evacuation or shelter protocols, and resource deployment strategies. A common error in planning is overreliance on written procedures without testing; therefore, plan maintenance includes regular review, exercises, and updates following lessons learned. Effective emergency plans are living documents shaped by continual feedback, ensuring readiness evolves alongside organizational and environmental change.

Incident Command Structure and Coordination

During an emergency, clarity of command determines the difference between chaos and control. The Incident Command System (ICS) provides a standardized management structure that organizes personnel, facilities, equipment, and communication under unified direction. It functions through modular expansion, adapting to event size and complexity. Command establishes strategic objectives, while sections for operations, planning, logistics, and finance manage specific functions. ICS emphasizes clear chain of command, common terminology, and integrated communication, enabling multiple agencies or departments to cooperate without conflict. For industrial facilities, ICS principles translate into site emergency command teams led by designated incident commanders who coordinate with public responders. The framework's scalability allows seamless integration with the National Incident Management System, ensuring compatibility with public-sector emergency operations. Practiced consistently, ICS instills discipline and predictability under stress, ensuring that decisions, information, and resources flow efficiently even in rapidly evolving conditions.

Communication, Information Management, and Public Coordination

Communication is the lifeline of emergency response. Reliable, redundant, and understandable channels ensure that instructions reach those who must act and that accurate information flows upward to decision-makers. Internal communication encompasses alarms, paging, radios, and digital platforms; external communication links the organization with first responders, regulators, and the community. NFPA 1600 emphasizes establishing protocols for message approval, public information dissemination, and rumor control. Information management also includes documentation—tracking incident status, decisions, and resource use to support accountability and post-event analysis. In modern systems, communication planning accounts for cyber resilience and interoperability among various technologies. Cultural and linguistic considerations further influence effectiveness; messages must be clear to diverse workforces and accessible to individuals with disabilities. Maintaining communication continuity ensures that even when power, networks, or physical facilities fail, essential coordination persists. Ultimately, transparent communication preserves trust and confidence, vital assets during any emergency.

Training, Exercises, and Capability Validation

Training transforms policy into practiced skill. Every employee must understand their role during an emergency, from initial notification to evacuation or response tasks. Training programs should be tiered according to responsibility levels—basic awareness for all personnel, functional instruction for response teams, and command-level decision training for leaders. NFPA 1600 and business continuity standards emphasize exercises as validation mechanisms: tabletop reviews, functional simulations, and full-scale drills test procedures under realistic conditions. These exercises reveal weaknesses not apparent on paper, allowing corrective actions before an actual event. Evaluation and after-action reviews document performance gaps and assign accountability for improvement. Equally important is repetition; capability decays without continuous reinforcement. Integrating training into orientation, refresher programs, and performance evaluations embeds preparedness into the organizational rhythm. Through structured and continuous education, emergency response evolves from theoretical compliance to a culture of competence.

Resource Management and Mutual Aid

Emergencies demand resources beyond normal operations—personnel, equipment, facilities, and information. Resource management involves pre-identifying, cataloging, and maintaining these assets so they are ready when needed. This includes inventories of critical supplies, vendor agreements for rapid resupply, and mutual aid arrangements with neighboring organizations or agencies. NFPA 1600 highlights the importance of maintaining redundancy and ensuring compatibility of equipment, such as communication gear or protective equipment, across partner entities. Financial resource management is also critical; documented authorization procedures prevent misuse while enabling rapid procurement during crises. Coordination with logistics and supply chain management ensures that emergency resources do not disrupt essential business functions. Well-structured resource systems balance flexibility and control—able to mobilize quickly without losing accountability. In practice, resource readiness reflects organizational maturity: the ability to sustain response operations over time without depleting normal operational capability.

Emergency Response Operations and Life Safety Priorities

When an incident occurs, immediate response focuses on life safety, incident stabilization, and property protection. Life safety encompasses evacuation, rescue, and medical response. Plans must define primary and secondary evacuation routes, assembly areas, and procedures for accounting for personnel. For facilities handling hazardous materials, specialized response teams trained in containment and decontamination ensure environmental protection alongside human safety. Incident stabilization requires controlling the hazard source—shutting down systems, activating suppression, or isolating affected areas. Coordination between on-site teams and external emergency services ensures that control efforts complement rather than conflict. Time-critical decisions depend on situational awareness maintained through continuous communication and monitoring. Post-response transition involves securing the scene for investigation and initiating recovery operations. By adhering to pre-established priorities, response teams maintain focus even under pressure, ensuring that every action supports the overarching goals of protecting people first and assets second.

Business Continuity and Organizational Resilience

Business continuity bridges immediate response with long-term recovery. It ensures that essential functions continue despite disruption and that restoration occurs efficiently. Continuity planning identifies critical processes, acceptable downtime, and resource dependencies. Strategies include alternate worksites, remote communication capability, data backups, and succession for leadership roles. Resilience, as described in the organizational continuity literature, extends beyond continuity—it represents the ability to adapt, absorb, and thrive after crisis. A resilient organization learns from adversity, integrating lessons into stronger design and culture. NFPA 1600 unites continuity and emergency management, emphasizing shared governance and mutual reinforcement. Effective continuity requires cross-functional collaboration between safety, information technology, human resources, and operations. Testing continuity plans through simulations confirms their practicality under stress. When resilience becomes embedded at every organizational level, emergencies cease to be existential threats and instead become tests of the system's design strength and adaptability.

Recovery, Restoration, and Post-Incident Evaluation

Recovery begins once the immediate hazard is controlled but continues long after physical repair. It encompasses restoration of operations, employee welfare, community relations, and reputation management. NFPA 1600 and continuity standards advocate structured recovery phases—short-term stabilization, intermediate restoration, and long-term redevelopment. Documentation from the response phase guides decision-making, while damage assessments determine priorities for repair and financial recovery. Coordination with insurers, regulators, and community stakeholders ensures transparency and compliance. Post-incident evaluations convert experience into improvement. After-action reports analyze performance, identify root causes, and assign corrective actions. These findings feed back into mitigation and preparedness planning, completing the emergency management cycle. Recovery also includes psychological support for affected personnel, acknowledging that resilience is as much human as structural. True organizational maturity is measured not by the absence of incidents but by the capacity to recover with integrity and insight.

Integration, Governance, and Continuous Improvement

Emergency preparedness succeeds only when integrated into the organization's governance structure. Leadership commitment, policy alignment, and performance measurement ensure that preparedness remains visible and prioritized. Integration links emergency management with risk management, occupational safety, security, and sustainability functions. Governance involves establishing metrics—response time, training compliance, and exercise performance—to evaluate program effectiveness. Continuous improvement relies on lessons learned, benchmarking against standards, and adapting to emerging threats such as cyber disruption or climate-driven hazards. NFPA 1600 concludes with a simple principle: preparedness is a living system. Its effectiveness depends not on perfection at one point in time but on the discipline of constant refinement. When leadership views preparedness as a strategic investment rather than a regulatory burden, the organization achieves lasting resilience—ready not only to respond, but to endure and evolve beyond crisis.

Chapter 6 – Industrial Hygiene and Occupational Health

The Science and Purpose of Industrial Hygiene

Industrial hygiene is the applied science devoted to anticipating, recognizing, evaluating, and controlling workplace conditions that may cause injury or illness. It functions as the bridge between occupational safety and public health, focusing on the human–environment interaction in workplaces. Industrial hygienists employ environmental monitoring and analytical techniques to detect exposure levels to physical, chemical, biological, or ergonomic hazards. The discipline integrates toxicology, physics, chemistry, and physiology to understand how environmental stressors affect health. Its purpose is preventive: to identify risks before harm occurs and to design systems that maintain worker well-being. The scope ranges from evaluating confined space atmospheres to measuring noise exposure or assessing ergonomic strain. The practice is both technical and ethical; it requires not only precision in measurement but also judgment in applying controls that respect worker diversity, environmental stewardship, and sustainability. Ultimately, industrial hygiene embodies the commitment to ensure that productivity does not come at the cost of health.

Anticipation, Recognition, Evaluation, and Control Framework

The foundation of industrial hygiene rests on a continuous four-step framework: anticipation, recognition, evaluation, and control. Anticipation involves foreseeing potential hazards through design reviews, chemical inventories, and task analyses before work begins. Recognition identifies the actual hazards present—chemical vapors, dusts, noise, or biological agents—through inspection and process knowledge. Evaluation quantifies these hazards, using air sampling, biological monitoring, and data comparison to exposure limits. Control completes the cycle by applying engineering, administrative, or personal protective measures to reduce risk. Each stage informs the next, creating a dynamic feedback loop. Anticipation without evaluation is speculation; evaluation without control is inaction. Effective industrial hygiene demands the practitioner to merge scientific rigor with practical field insight. As processes evolve, so must the cycle, ensuring the workplace remains adaptive to emerging technologies, materials, and health standards.

Chemical Hazards and Toxicological Principles

Chemical hazards are among the most pervasive challenges in occupational environments. They encompass gases, vapors, mists, dusts, fumes, and fibers, each capable of producing acute or chronic health effects depending on toxicity, dose, and exposure duration. Understanding toxicology—the study of how substances affect living organisms—is central to industrial hygiene practice. Toxic effects depend not only on chemical identity but also on physical form, route of entry, and individual susceptibility. Acute toxicity may cause immediate symptoms such as irritation or narcosis, while chronic exposure may result in long-term damage to organs or the nervous system. Concepts such as threshold limit values (TLVs), permissible exposure limits (PELs), and dose–response relationships allow quantitative comparison between measured exposures and acceptable standards. Industrial hygienists use these data to prioritize controls, design ventilation, and select protective equipment. Recognizing that toxicity defines the inherent capacity for harm while hazard represents the likelihood of that harm occurring under specific conditions allows professionals to focus prevention where exposure intersects with susceptibility.

Routes of Exposure and Dose Relationships

For any chemical or biological agent to exert an effect, it must enter the body and reach a site where interaction occurs. The principal routes of occupational exposure are inhalation, skin absorption, ingestion, and injection. Inhalation is the most common, as airborne contaminants—gases, vapors, dusts, or mists—penetrate deep into the respiratory system. Dermal absorption becomes significant with solvents, pesticides, or metals capable of permeating the skin. Ingestion often arises from poor hygiene practices, and injection risks occur during handling of sharp contaminated objects. The dose, defined as the amount entering the body, determines the magnitude of response. The relationship between dose and effect is rarely linear; individual variability, exposure duration, and cumulative burden shape outcomes. Time-weighted averages and short-term exposure limits account for fluctuating concentrations, offering a practical measure of cumulative risk. Recognizing and interrupting these exposure pathways through design, procedure, or protection is the essence of hygienic control.

Air Sampling and Exposure Monitoring

Air sampling transforms invisible risk into measurable data. Through collection and analysis, the hygienist establishes the presence and concentration of airborne contaminants. Methods include grab sampling for short-duration tasks and integrated sampling for full-shift exposure assessment. Modern practice employs calibrated pumps, sorbent tubes, filters, and direct-reading instruments such as photoionization detectors or infrared analyzers. Sampling design considers contaminant behavior, process variability, and worker mobility. Chain-of-custody protocols ensure legal defensibility, while calibration and documentation maintain data integrity. Results are interpreted against exposure limits—TWA, STEL, or ceiling values—set by authoritative bodies. Continuous monitoring systems extend this principle into real-time surveillance, offering early warning for process deviations. Air sampling is not an end in itself but a diagnostic tool guiding engineering design, verifying control effectiveness, and supporting regulatory compliance. Its success depends on both technical accuracy and the professional's understanding of context, ensuring data represent true exposure rather than isolated measurement.

Biological and Physical Hazards

Beyond chemicals, workers encounter biological agents and physical stressors that threaten health. Biological hazards include bacteria, fungi, viruses, and allergens capable of causing infection, hypersensitivity, or toxicity. Environments such as healthcare facilities, agriculture, and waste management are particularly vulnerable. Recognition involves assessing sources, transmission routes, and environmental conditions supporting growth. Control emphasizes sanitation, ventilation, and vaccination programs. Physical hazards encompass noise, vibration, radiation, temperature extremes, and pressure changes. Each presents measurable stress on the body—hearing loss from prolonged noise, thermal strain from heat, or musculoskeletal fatigue from vibration. Assessment employs instruments such as dosimeters, radiometers, and heat stress monitors. The hygienist integrates these data with physiological response to establish safe exposure conditions. The holistic approach of industrial hygiene demands equal attention to biological and physical factors, acknowledging that human health is influenced by the combined load of all environmental stressors.

Exposure Limits and Evaluation Methods

Exposure limits represent professional consensus on safe concentrations under typical working conditions. Regulatory agencies and scientific organizations—such as OSHA, NIOSH, and ACGIH—publish values including permissible exposure limits, recommended exposure limits, and threshold limit values.

These may be expressed as time-weighted averages, short-term exposure limits, or ceiling concentrations. Evaluation involves comparing measured data with these benchmarks, interpreting variance, and considering uncertainty. Biological exposure indices complement air sampling by measuring internal dose through blood, urine, or exhaled breath analysis. The hygienist must also account for mixed exposures, synergistic effects, and vulnerable populations. Data interpretation moves beyond compliance toward risk characterization, determining whether exposures are as low as reasonably achievable. Evaluation culminates in communication—translating technical findings into actionable insight for decision-makers, supervisors, and workers.

Control Strategies and the Hierarchy of Controls

Controlling occupational hazards follows a structured hierarchy emphasizing prevention over personal protection. The sequence begins with elimination—removing the hazard entirely—followed by substitution with less hazardous materials or processes. Engineering controls, such as enclosure and local exhaust ventilation, isolate hazards from workers. Administrative controls modify behavior through scheduling, rotation, or procedural discipline. Personal protective equipment (PPE) serves as the final defense, effective only when properly selected, fitted, and maintained. The hierarchy reflects both effectiveness and reliability: engineering changes endure longer than behavioral measures. Modern industrial hygiene integrates control verification, ensuring that implemented measures achieve intended reductions through follow-up monitoring. In practice, multiple controls often operate concurrently, forming redundant protection layers. The professional's responsibility is to balance practicality with protection, achieving safety without compromising productivity or comfort.

Occupational Health Surveillance and Medical Interface

Industrial hygiene data attain full value when linked with occupational health surveillance. Monitoring exposure trends alongside medical outcomes identifies emerging risks and validates control efficacy. Surveillance may include routine health examinations, biological monitoring, and symptom tracking for populations with specific exposures. The collaboration between hygienists and occupational health professionals enables early detection of conditions such as hearing loss, respiratory disease, or dermatitis. Confidentiality, informed consent, and ethical standards govern all data handling. When abnormal trends appear, investigation traces causation through environmental and process review. Integration of exposure records with medical findings supports epidemiologic analysis, strengthening evidence for regulatory and preventive action. Occupational health surveillance thus transforms measurement into protection, ensuring that data translate directly into improved well-being.

Indoor Air Quality and Ergonomic Considerations

As industrial operations extend into office and institutional settings, indoor air quality and ergonomics have become central to worker health. Poor ventilation, inadequate filtration, and contaminant buildup contribute to symptoms collectively termed "sick building syndrome." Evaluation involves measuring carbon dioxide as a ventilation indicator, assessing particulate and microbial levels, and reviewing HVAC performance. Ergonomic factors—workstation design, repetitive motion, and postural strain—represent another dimension of occupational health. Industrial hygienists collaborate with engineers and health specialists to design environments that accommodate human variability. Solutions include adjustable workstations, improved lighting, and microclimate control. Attention to indoor air and ergonomics aligns with the preventive philosophy of industrial hygiene: to design systems that fit human needs rather than force adaptation to poorly designed spaces.

Program Management, Documentation, and Continuous Improvement

Sustaining occupational health requires structured program management. Documentation, quality assurance, and continuous improvement ensure that industrial hygiene functions as an integral part of organizational systems. Records of sampling, calibration, control implementation, and training form the evidence base for regulatory compliance and internal review. Audits and periodic program evaluations identify gaps, prompting corrective and preventive action. Continuous improvement aligns with management standards such as ISO 45001, emphasizing worker participation and leadership commitment. As new technologies and materials emerge—nanoparticles, additive manufacturing, bioaerosols—industrial hygiene adapts by expanding methodologies and refining exposure models. The profession's endurance lies in its scientific integrity and its unwavering focus on human health, ensuring that every innovation in industry is matched by innovation in protection.

Chapter 7 – Environmental Management and Sustainability

Foundations of Environmental Management

Environmental management represents the structured coordination of policies, procedures, and technologies designed to reduce the ecological impact of human activity. Its origins lie in the growing recognition that industrial and urban development must coexist with environmental integrity. Within professional safety and health practice, environmental management bridges engineering, regulatory compliance, and sustainability. The foundation of modern systems, such as ISO 14001, is the continuous improvement cycle—plan, implement, evaluate, and act. This framework ensures that organizations identify environmental aspects, evaluate significance, and implement controls that prevent pollution at the source. Environmental management integrates with quality and occupational health systems, forming a triad that governs risk in all its forms. The discipline is preventive rather than reactive, emphasizing anticipation of environmental harm before it occurs. The goal is not only compliance with environmental law but stewardship that sustains resources for future generations, aligning operational success with ecological responsibility.

Environmental Engineering and Scientific Principles

At its core, environmental management draws heavily on the science of environmental engineering. This field applies physics, chemistry, and biology to understand and control pollution in air, water, and soil systems. Environmental engineers analyze mass and energy balances, transport phenomena, and reaction kinetics to design systems that mitigate contamination. Understanding pollutant pathways—dispersion in air, advection and diffusion in water, and adsorption in soils—enables prediction of environmental behavior and development of control technologies. In water management, treatment processes remove solids, pathogens, and chemicals through sedimentation, filtration, and biological oxidation. Air pollution control employs capture, scrubbing, and catalytic conversion to meet emission limits. Waste management strategies emphasize reduction, reuse, and recycling before resorting to disposal. The integration of these engineering principles with management systems ensures that prevention and control operate simultaneously. Through science-based decision-making, environmental management transcends compliance, advancing toward restoration and resilience of natural systems.

Pollution Prevention and Control Strategies

Pollution prevention is the central pillar of sustainable environmental management. It focuses on eliminating or reducing contaminants at the source rather than treating them after release. Strategies include process redesign, material substitution, and efficiency improvement to minimize waste generation. For instance, closed-loop systems in manufacturing recycle water and recover solvents, reducing both consumption and discharge. Air emission controls such as fabric filters, scrubbers, and electrostatic precipitators prevent particulate and gaseous pollutants from escaping into the atmosphere. Wastewater treatment systems integrate mechanical, biological, and chemical processes to meet stringent effluent standards. Solid waste management applies hierarchy principles—reduce, reuse, recycle, and recover—before disposal. Pollution prevention extends beyond equipment to include operational discipline, maintenance reliability, and employee awareness. The overarching principle is that prevention costs less than remediation. By integrating control technologies with management oversight, organizations maintain cleaner production and reduce liability, while protecting communities and ecosystems.

Air Quality and Atmospheric Protection

Air quality management addresses pollutants that affect both human health and climate stability. Industrial processes emit particulate matter, sulfur dioxide, nitrogen oxides, carbon monoxide, and volatile organic compounds, each governed by distinct chemical transformations in the atmosphere. Environmental engineers analyze stack emissions, dispersion patterns, and chemical reactions that lead to smog or acid rain formation. Control technologies include thermal oxidizers for organic vapors, low-NOx burners for combustion sources, and filtration systems for dust control. Monitoring plays a crucial role; continuous emission systems verify compliance and support transparent reporting. Beyond regulatory control, air quality management now encompasses greenhouse gas reduction under global sustainability frameworks. Carbon footprint assessments quantify direct and indirect emissions, guiding mitigation strategies such as energy efficiency and renewable integration. Maintaining air quality is both a local health imperative and a global responsibility, demonstrating the interdependence of industrial activity and planetary systems.

Water Resources and Wastewater Management

Water is both a vital resource and a vulnerable medium for pollution. Industrial operations may alter its quantity, quality, or ecological function if mismanaged. Effective water management begins with source protection, ensuring that withdrawal does not exceed renewal capacity. Wastewater management treats used water before discharge to remove contaminants that could harm aquatic life or human users. Treatment trains typically include preliminary screening, primary sedimentation, secondary biological oxidation, and advanced polishing such as filtration or disinfection. Emerging concerns such as microplastics, pharmaceutical residues, and nutrient overloads challenge traditional systems, requiring innovation in advanced oxidation and membrane technologies. The guiding philosophy mirrors that of occupational hygiene—measure, control, and verify. By minimizing consumption through recycling and reuse, industries reduce both environmental impact and operational cost. Comprehensive water management thus supports sustainability goals, preserving ecological balance while ensuring operational continuity.

Waste Management and Resource Recovery

Waste management has evolved from disposal toward circular resource recovery. The modern view treats waste as misplaced material value rather than unavoidable residue. Effective systems classify wastes by hazard, composition, and recoverability. Hazardous waste management applies cradle-to-grave tracking, ensuring proper containment, transport, and disposal in compliance with environmental regulations. Non-hazardous materials such as paper, metals, and plastics enter recycling streams that conserve raw materials and energy. Industrial ecology extends this principle further, designing processes where waste from one system becomes input for another. Energy recovery through incineration, anaerobic digestion, or biogas production converts waste into useful power while minimizing landfill dependency. Landfills themselves have become engineered containment systems equipped with liners, leachate collection, and gas recovery. The transition from linear to circular economy represents a cultural shift—from managing waste to managing materials—anchoring sustainability in practical resource efficiency.

Environmental Management Systems and Auditing

An Environmental Management System (EMS) structures environmental protection into a formalized, auditable process. The ISO 14001 standard provides the global benchmark, establishing requirements for policy, planning, implementation, and continual improvement. An EMS begins with identifying environmental aspects—activities that interact with the environment—and evaluating their significance. Objectives and targets translate this analysis into measurable performance commitments. Implementation involves defining responsibilities, training personnel, and maintaining operational controls. Regular audits verify compliance and uncover opportunities for improvement. Documentation ensures traceability and accountability, while management reviews sustain leadership involvement. Integration with other systems such as ISO 9001 and ISO 45001 ensures consistency across quality, safety, and environmental domains. The outcome is predictability: environmental performance becomes systematic rather than incidental. Through disciplined management, organizations move beyond compliance toward excellence, embedding environmental consideration into every business decision.

Sustainability and Environmental Ethics

Sustainability extends the scope of environmental management from control to conscience. It encompasses meeting current needs without compromising the ability of future generations to meet theirs. Environmental ethics demands that organizations evaluate not only economic profit but ecological and social impact. The sustainability framework—often expressed through the "triple bottom line" of people, planet, and profit—redefines success as balanced performance across all three. Decisions about energy use, material sourcing, and waste generation are assessed for long-term consequences. Transparency in reporting and stakeholder engagement ensures accountability. Ethical management also recognizes environmental justice, ensuring that protection extends to communities historically burdened by industrial development. Sustainability transforms compliance into leadership, positioning organizations as responsible stewards rather than reluctant participants. When ethical awareness informs operational strategy, environmental performance becomes intrinsic to organizational identity.

Environmental and Social Governance (ESG) Integration

Environmental and Social Governance, or ESG, has emerged as the leading framework for translating sustainability into measurable, reportable performance. ESG integrates environmental metrics—such as carbon intensity, water efficiency, and waste reduction—with social and governance criteria encompassing worker welfare, community relations, and ethical oversight. Investors and stakeholders increasingly evaluate ESG performance as a proxy for organizational resilience and long-term value. Implementing ESG requires cross-functional collaboration among environmental engineers, safety professionals, and financial officers. Data integrity and transparency are paramount; metrics must withstand public scrutiny and align with recognized standards. ESG reporting thus extends beyond environmental compliance into corporate strategy, influencing investment, procurement, and risk management. Properly executed, ESG integration strengthens credibility and competitiveness, illustrating that environmental stewardship and economic performance are mutually reinforcing rather than conflicting objectives.

Life Cycle Assessment and Sustainable Design

Life Cycle Assessment (LCA) provides the analytical backbone of sustainable design. It evaluates environmental impact from raw material extraction through production, use, and ultimate disposal. By quantifying energy and material flows, LCA identifies hotspots where design or process changes yield the greatest benefit. In engineering, this translates into selecting recyclable materials, optimizing product durability, and reducing embodied energy. Sustainable design extends these insights into architecture, manufacturing, and product development, promoting systems that conserve resources throughout their life span. LCA findings also support transparent labeling and environmental declarations, allowing consumers and stakeholders to make informed decisions. When integrated with management systems and ESG frameworks, life cycle thinking ensures that sustainability permeates every design stage. The outcome is a shift from reactive mitigation to proactive innovation, where products and processes are conceived for minimal impact from the outset.

Climate Change, Resilience, and Continuous Improvement

Climate change has redefined the context of environmental management. Rising temperatures, extreme weather, and shifting resource availability challenge the stability of industrial systems. Environmental management must therefore include adaptation and resilience planning—evaluating vulnerabilities and fortifying infrastructure against foreseeable disruptions. Carbon management, energy transition, and renewable integration become central strategies for both mitigation and adaptation. Continuous improvement links these initiatives into a feedback loop of measurement, action, and innovation. Through environmental auditing, performance data inform new targets, while technology advances expand possibilities for emission reduction and resource conservation. The goal is perpetual progress rather than static compliance. Resilient organizations anticipate change, learn from experience, and transform environmental responsibility into a source of strength. In this evolving landscape, sustainability ceases to be an external requirement and becomes the defining measure of organizational maturity and foresight.

Chapter 8 – Training, Education, and Communication

Foundations of Learning and Professional Development

Training and education form the connective tissue between technical knowledge and safe behavior. They transform policy into practice and theory into performance. The foundation of learning in occupational safety rests on understanding how adults acquire, retain, and apply knowledge. Adult learners bring experience, expectations, and self-direction to the classroom; therefore, effective training must respect their autonomy and relevance needs. The learning process engages three domains—cognitive, psychomotor, and affective—each essential for building competence in safety-critical work. Education focuses on understanding principles, while training emphasizes applying them under realistic conditions. Together, they develop both awareness and action. Successful learning programs rely on repetition, feedback, and reflection. They are designed not merely to transfer information but to change behavior in ways that prevent incidents. In professional safety practice, training is not an isolated event but an integral component of continuous improvement, reinforcing a culture where knowledge and vigilance sustain one another.

Principles of Adult Learning and Instructional Design

Adult learning theory asserts that experience is the foundation for new understanding. Learners are motivated when they see direct relevance to their tasks and when they can control their learning pace. Instructional design therefore begins with identifying learning needs and translating them into measurable objectives. The process follows a systematic model—analyzing needs, designing structure, developing materials, implementing delivery, and evaluating results. Training must connect conceptual knowledge to authentic job contexts through examples, simulations, and discussion. The educator becomes a facilitator rather than a lecturer, guiding learners to build insight through problem-solving and feedback. Blended approaches—combining classroom, online, and hands-on learning—accommodate diverse preferences and reinforce retention. Evaluation of comprehension through observation or performance demonstration ensures transfer to the workplace. By aligning learning design with adult learning principles, organizations move beyond compliance training toward true capacity building, where knowledge translates into safer and more consistent performance.

Developing Training Needs Assessment and Objectives

Effective training begins with a precise understanding of what must be learned and why. A training needs assessment identifies gaps between existing and desired performance, considering regulatory obligations, technological change, and organizational priorities. Sources include accident investigations, audits, and worker feedback, all of which reveal patterns of misunderstanding or procedural weakness. Once gaps are defined, clear objectives articulate the desired outcome—what learners should know, do, or value upon completion. Objectives guide content selection, instructional strategy, and assessment method, ensuring coherence between purpose and practice. Each objective must be specific, measurable, achievable, relevant, and time-bound. In the safety field, well-constructed objectives convert complex regulations into actionable behavior, such as demonstrating correct lockout procedures or interpreting hazard communication labels. Needs assessment also identifies barriers—language, literacy, or cultural differences—that may impede learning. Addressing these ensures equity and effectiveness. A systematic assessment process guarantees that training investments yield measurable safety improvement rather than unstructured activity.

Instructional Delivery and Engagement Strategies

Delivering training effectively requires more than subject expertise; it demands skill in engagement. Learners retain information best when they are active participants, not passive recipients. Effective instructors combine verbal clarity with visual reinforcement, demonstration, and interaction. Storytelling and case analysis bridge theory and practice, allowing participants to connect lessons to real experiences. In technical safety training, demonstration and practice sessions under supervision transform procedural knowledge into physical skill. Group discussions foster peer learning, while questioning techniques stimulate critical thinking. Visual aids such as diagrams, simulations, and videos support comprehension of complex concepts like hazard recognition or emergency response sequences. Online and virtual platforms extend reach, but they must still maintain interaction through assessments and feedback loops. Regardless of delivery mode, the essence of engagement lies in creating an environment where learners feel respected, challenged, and supported—conditions that convert attention into understanding and understanding into consistent safe behavior.

Evaluating Training Effectiveness and Performance Outcomes

Evaluation confirms whether training has achieved its intended results. The classical approach, grounded in behavioral science, measures effectiveness across multiple levels: reaction, learning, behavior, and results. Reaction gauges participant satisfaction, learning measures knowledge gain, behavior evaluates on-the-job application, and results assess impact on organizational performance. Evaluation tools include pre- and post-tests, observation checklists, performance metrics, and safety incident data. Linking outcomes to measurable performance ensures accountability. When evaluation reveals gaps, training design must be adjusted to strengthen delivery or content. Effective evaluation also recognizes that knowledge retention diminishes over time; periodic refresher courses and reinforcement through supervision sustain learning. Integration with performance management systems ensures that training outcomes influence promotion, recognition, and corrective action. Ultimately, evaluation transforms training from a cost center into an investment in human capital, where measurable improvement in safety performance justifies continued organizational commitment.

Safety Communication and Information Flow

Communication in safety management extends beyond the classroom into the daily fabric of organizational life. It encompasses formal instruction, procedural documentation, signage, and informal dialogue between supervisors and workers. Effective communication ensures that information flows freely, accurately, and promptly across all levels. It must be clear, concise, and consistent, using language appropriate to the audience's literacy and cultural background. Visual communication through standardized symbols and color coding reinforces understanding even where language barriers exist. Regular safety meetings, toolbox talks, and digital updates sustain awareness. Equally critical is upward communication, allowing workers to report hazards, near misses, and improvement ideas without fear of reprisal. When communication is bidirectional, it becomes a tool for learning rather than control. Transparent information flow cultivates trust, enabling rapid response to emerging risks and embedding safety as a shared organizational language.

Building and Sustaining a Safety Culture

Training and communication together shape the culture in which safety decisions are made. Safety culture represents the collective values, beliefs, and behaviors that determine how safety is prioritized in everyday operations. A strong culture emerges when leadership demonstrates visible commitment and employees internalize safety as personal responsibility. Cultural reinforcement occurs through consistent messaging, recognition of safe behavior, and accountability for unsafe actions. Training contributes by transmitting not only procedures but values—why safety matters and how each individual influences outcomes. Measurement through surveys, observations, and incident trends provides insight into cultural maturity. When culture is robust, compliance becomes natural; safety is practiced because it is valued, not merely required. Maintaining this culture requires continuous dialogue, reinforcement, and leadership modeling, ensuring that the organization's core identity aligns with its safety aspirations.

Human Performance and Behavioral Reliability

Human performance improvement focuses on understanding why people act as they do within their working environment. It recognizes that most errors are system-induced rather than purely individual. By analyzing context—procedures, tools, environment, and feedback—organizations can design conditions that make safe behavior the easiest behavior. Training supports this by equipping workers with knowledge of human limitations and by embedding error prevention techniques such as self-checks, peer verification, and situational awareness. Behavioral reliability improves when systems reward correct actions and provide timely feedback for correction. Observation-based safety programs, when properly implemented, convert peer observation into learning rather than policing. Human performance frameworks treat errors as information rather than failure, turning incidents into opportunities for systemic improvement. Integrating these concepts into training ensures that human factors are addressed as part of design, supervision, and communication, fostering operational reliability through human strength rather than fear of error.

Diversity, Accessibility, and Cross-Cultural Training

Modern workplaces are increasingly diverse in language, culture, and physical ability. Training programs must therefore be inclusive by design. Content should respect cultural differences, avoid assumptions about prior knowledge, and use examples relevant to the participants' experience. Accessibility standards require materials and delivery methods that accommodate disabilities—captioned videos, adaptive devices, and clear visual aids. Multilingual resources and interpreters bridge linguistic barriers, while culturally aware instruction builds mutual respect. In safety-critical training, inclusivity is not only ethical but practical; misunderstanding instructions can result in injury or loss. Involving representatives from diverse groups in training design and feedback ensures authenticity and trust. Inclusive training recognizes that diversity enriches learning and strengthens organizational resilience. When every individual can access and apply knowledge effectively, safety performance becomes equitable and universally understood.

Continuous Improvement and Learning Organization

The hallmark of a mature safety program is its transformation into a learning organization. Continuous improvement in training and communication mirrors the broader quality cycle—plan, do, check, act. Lessons from incidents, audits, and evaluations feed back into revised content and improved delivery methods. Technology enables adaptive learning platforms that adjust difficulty based on performance, ensuring ongoing engagement. Knowledge management systems capture institutional experience, preserving learning across personnel changes. Leadership plays a pivotal role by modeling learning behavior—acknowledging errors, encouraging inquiry, and rewarding innovation. When organizations view every event, from success to failure, as data for improvement, they achieve resilience. Training thus evolves from a periodic requirement to a perpetual process of refinement. The result is an organization where safety, communication, and learning form a continuous cycle of growth, sustaining excellence through shared knowledge and adaptive capability.

Chapter 9 – Environmental Health and Safety Program Management and Leadership

Foundations of Safety Program Management

Safety management is the deliberate orchestration of policies, processes, and behaviors to prevent harm and ensure regulatory compliance. At its foundation lies a structured system that converts organizational values into daily practice. Environmental health and safety (EHS) management integrates planning, execution, and evaluation—each cycle ensuring continual improvement. A successful program begins with leadership commitment, clearly defined responsibilities, and measurable objectives. Strategic alignment with corporate goals transforms safety from a compliance cost into a productivity driver. The management function also includes monitoring, documentation, and corrective actions to maintain consistency and accountability. Through proactive planning, incident investigation, and periodic review, EHS programs evolve dynamically rather than reactively. The ultimate goal is not merely to meet legal standards but to embed safety into the organizational culture as a shared moral and operational priority. Sound management converts obligation into opportunity, shaping workplaces that are both safe and sustainable.

Legal and Regulatory Framework for EHS Leadership

Every EHS program operates within a complex web of statutory obligations and case law precedents. Understanding these legal frameworks is essential for effective leadership. The Occupational Safety and Health Act (OSH Act) establishes the foundation for federal workplace safety regulation, granting the Occupational Safety and Health Administration authority to issue and enforce standards. Complementary state programs may operate under OSHA-approved plans, adapting requirements to local conditions. Beyond OSHA, other federal laws—such as the Clean Air Act, Clean Water Act, and Resource Conservation and Recovery Act—shape environmental aspects of management. Legal obligations also extend to worker compensation statutes and tort liability. Leadership requires awareness that compliance does not end with written procedures; it involves demonstrable due diligence, training, and enforcement. The legal doctrine of "reasonable care" holds organizations accountable for foreseeable risks, while evolving case law continually refines expectations. EHS leaders must therefore stay informed, ensuring that their decisions anticipate legal scrutiny and align with both statutory and ethical responsibility.

Governance, Accountability, and Ethical Responsibility

EHS governance extends beyond compliance to encompass the ethical stewardship of human and environmental well-being. Accountability systems ensure that safety obligations are owned at every organizational level. Governance frameworks define authority, responsibility, and oversight mechanisms that guarantee transparency and performance integrity. Ethical responsibility, though less tangible than regulation, provides the moral foundation of safety leadership. It demands that decisions consider their broader consequences for workers, communities, and ecosystems. Ethical governance also ensures fairness—applying rules consistently and protecting the rights of employees to safe conditions. Accountability is reinforced through internal audits, performance reviews, and corrective action systems that reward compliance and address lapses promptly. Leaders who model ethical behavior set the tone for the entire organization; their integrity fosters trust and voluntary compliance. Governance in EHS is thus both structural and cultural—a system of controls built upon values that affirm the intrinsic worth of human life and environmental balance.

Strategic Planning and Policy Integration

Strategic safety planning transforms vision into measurable progress. It aligns EHS goals with the organization's overall mission and performance objectives. The process begins with defining a clear policy statement expressing commitment to health, safety, and environmental stewardship. Strategic plans identify key risks, prioritize interventions, and allocate resources efficiently. Integration across departments ensures that safety considerations influence decisions in engineering, procurement, human resources, and operations. Periodic reviews and measurable milestones support continuous improvement, while cross-functional collaboration encourages shared ownership. Effective planning also requires forecasting emerging risks such as technological change, supply chain vulnerabilities, or evolving regulations. By embedding safety into corporate strategy, leaders ensure that risk management and productivity coexist rather than compete. Strategic EHS planning thus moves from reactive compliance toward proactive innovation, positioning the organization as a responsible, resilient enterprise prepared for future challenges.

Leadership Commitment and Organizational Influence

Leadership commitment is the single most decisive factor in EHS performance. When executives prioritize safety through visible actions—such as resource allocation, participation in audits, and communication—employees perceive safety as integral to organizational success. Commitment is demonstrated through example rather than proclamation. Leaders influence safety culture by modeling adherence to procedures, valuing employee input, and addressing hazards promptly. This influence cascades downward, shaping behaviors and expectations throughout the hierarchy. Leadership training in communication, motivation, and ethical reasoning enhances consistency between words and actions. When safety metrics are included in executive performance evaluations, accountability becomes institutionalized. True leadership goes beyond compliance; it inspires voluntary adherence to principles of care and prevention. By reinforcing that safety is inseparable from operational excellence, EHS leaders create environments where compliance is a natural expression of shared purpose rather than a forced obligation.

Performance Measurement and Continuous Improvement

Measurement is the feedback system that drives improvement. Performance indicators in EHS management fall into two broad categories: lagging indicators, which measure outcomes such as incident rates or regulatory citations, and leading indicators, which monitor proactive efforts such as inspections, training completion, and near-miss reporting. Balanced scorecards combine both, offering a comprehensive view of program health. Continuous improvement requires not only measurement but interpretation—understanding what data reveal about behavior, culture, and system effectiveness. Benchmarking against industry standards provides external validation, while root-cause analysis translates findings into preventive strategies. The Plan-Do-Check-Act cycle ensures that insights from audits and investigations inform policy revision and resource allocation. Continuous improvement transforms compliance into innovation, ensuring that safety performance evolves alongside operational complexity. The objective is not perfection but progress—a relentless pursuit of fewer incidents, stronger engagement, and deeper understanding of risk dynamics.

Incident Investigation and Corrective Action

Incident investigation is both a diagnostic and preventive tool. Its purpose extends beyond assigning blame to uncover the systemic factors that allowed the event to occur. Effective investigations employ structured methodologies—such as causal tree analysis or the "five whys" approach—to trace events back to root causes. Data collection includes witness statements, physical evidence, and procedural reviews, all conducted with objectivity and confidentiality. Corrective actions follow two levels: immediate controls to prevent recurrence and long-term system modifications to address underlying weaknesses. Tracking and verifying corrective action completion closes the feedback loop. Investigations also contribute to organizational learning, producing insights that inform future training and policy refinement. The goal is to convert every incident into an opportunity for improvement. Transparent communication of lessons learned reinforces trust and engagement, demonstrating that the organization values knowledge over punishment and prevention over reaction.

Risk Governance and Decision-Making

Risk governance is the disciplined management of uncertainty through informed decision-making. It encompasses identifying, assessing, and prioritizing risks, then allocating resources proportionate to their potential impact. Governance frameworks ensure that risk decisions are transparent, evidence-based, and aligned with organizational objectives. Decision-making integrates technical data with ethical and social considerations, recognizing that not all risks are purely quantitative. In safety management, the principle of "as low as reasonably achievable" guides risk tolerance, balancing practicality with protection. Leadership roles include setting acceptable risk thresholds, ensuring compliance with regulatory limits, and approving mitigation strategies. Modern governance also addresses emerging risks such as cybersecurity in safety systems or climate-related disruptions. By institutionalizing risk management into corporate governance, organizations achieve resilience—capable of adapting to uncertainty without compromising integrity or safety performance.

Communication, Engagement, and Stakeholder Trust

Trust is the currency of effective safety management. Transparent communication builds that trust among employees, regulators, and the public. EHS leaders must communicate complex information in ways that are accurate, accessible, and actionable. Engagement extends beyond informing to involving stakeholders in decision-making processes—through committees, consultations, and open forums. When workers participate in identifying hazards and designing solutions, compliance shifts from obligation to ownership. Externally, communicating environmental and safety performance through sustainability reports or public disclosures enhances credibility. The consistency and honesty of communication are vital, especially during crises, when uncertainty and fear amplify misinformation. A culture of open communication strengthens resilience by ensuring that critical information flows rapidly to those who need it most. Ultimately, stakeholder engagement transforms EHS management from a regulatory requirement into a collaborative social contract built on transparency and mutual respect.

Legal Preparedness and Compliance Management

Preparedness for legal scrutiny is integral to EHS leadership. Compliance management involves establishing systems that not only meet statutory requirements but document and demonstrate adherence. This includes maintaining up-to-date records, training logs, and audit reports that serve as evidence of due diligence. Legal preparedness also means anticipating potential litigation following incidents and ensuring that documentation, chain-of-custody, and communication protocols are defensible. The integration of legal awareness into management systems reduces vulnerability by ensuring decisions are both technically and legally sound. Collaboration with legal counsel strengthens interpretation of complex regulations and ensures that corrective actions address not only safety outcomes but compliance expectations. In an era of increasing enforcement and public accountability, preparedness is protection. The legally aware safety professional becomes an indispensable advisor, capable of navigating the intersection of law, ethics, and operational practice.

Integrating Leadership, Law, and Culture in EHS Systems

The highest expression of EHS maturity is the seamless integration of leadership, law, and culture. Leadership provides direction, law establishes boundaries, and culture sustains behavior. When these three converge, organizations achieve sustainable excellence. The legal system defines minimum expectations through statutes and case law, but leadership transforms these mandates into guiding principles that define identity. Culture then embeds them into the daily habits of every worker. Integration requires continuous reinforcement through education, recognition, and accountability. Leaders who understand both the technical and legal dimensions of safety foster an environment of informed empowerment—where compliance is a natural outcome of shared values. Ultimately, the fusion of law, management, and culture defines the modern EHS paradigm: one that safeguards people, preserves the environment, and exemplifies ethical governance in every decision.

Chapter 10 – Practice Questions

Questions (1-100)

Q1. A worker is exposed to 90 dB of noise for 4 hours and 85 dB for another 4 hours. What is the equivalent continuous noise level (Leq)?

A. 86 dB

B. 87 dB

C. 88 dB

D. 89 dB

Q2. A 200 kg load is suspended by two slings at equal angles of 60° from the horizontal. What is the tension in each sling?

A. 100 kg

B. 115 kg

C. 173 kg

D. 200 kg

Q3. A fall protection system has a free-fall distance of 6 ft and a deceleration distance of 3.5 ft. What is the total fall clearance required including 2 ft safety margin?

A. 9 ft

B. 11.5 ft

C. 13.5 ft

D. 15.5 ft

Q4. Calculate the incidence rate (IR) if there were 4 recordable cases and 500,000 employee hours worked.

A. 1.2

B. 1.6

C. 2.0

D. 3.0

Q5. A ventilation system provides 1,500 CFM to a room of 10,000 ft³. What is the air change rate per hour (ACH)?

A. 6

B. 8

C. 9

D. 12

Q6. Calculate Time-Weighted Average (TWA) exposure: 60 ppm for 2 h, 90 ppm for 4 h, 30 ppm for 2 h.

A. 45 ppm

B. 60 ppm

C. 65 ppm

D. 75 ppm

Q7. A trench is 8 ft deep in Type B soil. According to OSHA, the maximum allowable slope angle is approximately:

A. 34°

B. 45°

C. 53°

D. 60°

Q8. A process safety audit finds an annual cost of incidents at $25,000 and a total cost of risk at $250,000. What is the cost of incidents as a percentage of total risk cost?

A. 5%

B. 8%

C. 10%
D. 12%

Q9. What is the total force on a 10 m² wall from wind pressure of 500 Pa?
A. 2,000 N
B. 3,000 N
C. 5,000 N
D. 10,000 N

Q10. A sample has data points: 4, 8, 12, 16. What is the standard deviation?
A. 4.0
B. 4.6
C. 5.0
D. 5.3

Q11. If a pipeline carries 120 liters per minute of water, what is the flow rate in cubic meters per hour?
A. 4.2
B. 5.8
C. 7.2
D. 8.0

Q12. A 10 m ladder leans against a wall at a 70° angle. What is the horizontal distance from the wall?
A. 3.4 m
B. 5.0 m
C. 6.2 m
D. 7.1 m

Q13. A chemical reaction releases 500 kJ of heat in 20 seconds. What is the power output?
A. 10 kW
B. 20 kW
C. 25 kW
D. 30 kW

Q14. What is the resultant of two perpendicular forces of 30 N and 40 N?
A. 50 N
B. 60 N
C. 65 N
D. 70 N

Q15. A gas cylinder contains 0.5 m³ of gas at 200 bar. If released to 1 bar, what will be the volume (ideal behavior)?
A. 50 m³
B. 75 m³
C. 100 m³
D. 200 m³

Q16. An employee's exposure to a solvent is 300 ppm for 1 hour, 150 ppm for 3 hours, and 50 ppm for 4 hours. What is the 8-hour TWA?
A. 100 ppm
B. 112 ppm
C. 125 ppm
D. 150 ppm

Q17. A confined space has a volume of 40 m³. How long will it take to ventilate completely with 2 m³/min airflow?
A. 10 min
B. 15 min
C. 20 min

D. 25 min

Q18. A 12-inch pipe carries air at 1000 fpm. What is the airflow rate in CFM?

A. 650

B. 785

C. 940

D. 1,130

Q19. What is the gravitational potential energy of a 10 kg object raised 5 m above ground?

A. 25 J

B. 250 J

C. 490 J

D. 500 J

Q20. A process pump delivers 10 liters per second at 3 bar. What is the power required if efficiency is 80%?

A. 0.25 kW

B. 0.28 kW

C. 0.35 kW

D. 0.38 kW

Q21. A load of 600 kg is lifted vertically at a constant speed using a pulley system with 75% efficiency. What is the required input force if the mechanical advantage is 4?

A. 180 kgf

B. 200 kgf

C. 220 kgf

D. 230 kgf

Q22. If the force constant (k) of a spring is 500 N/m, how much potential energy is stored when it is stretched 0.2 m?

A. 5 J

B. 10 J

C. 20 J

D. 25 J

Q23. A 2 m^3 tank contains compressed air at 10 bar. When released to 1 bar, what volume will it occupy?

A. 15 m^3

B. 18 m^3

C. 20 m^3

D. 22 m^3

Q24. What is the kinetic energy of a 2 kg object moving at 5 m/s?

A. 10 J

B. 15 J

C. 20 J

D. 25 J

Q25. A pump discharges 5 liters/second of water through a 10 m elevation. What hydraulic power (in kW) is required? (Use g = 9.81 m/s^2)

A. 0.3 kW

B. 0.4 kW

C. 0.5 kW

D. 0.6 kW

Q26. If the temperature increases from 20°C to 80°C, what is the change in temperature on the Fahrenheit scale?

A. 90°F

B. 100°F

C. 108°F

D. 112°F

Q27. What is the density of a material weighing 25 kg and occupying 0.02 m^3?

A. 1,000 kg/m^3

B. 1,200 kg/m^3

C. 1,250 kg/m^3

D. 1,400 kg/m^3

Q28. An air sample at 1 atm and 25°C occupies 24 L. What will its volume be at 50°C and constant pressure?

A. 25 L

B. 26 L

C. 27 L

D. 28 L

Q29. A noise exposure of 95 dBA is measured for 2 hours. According to OSHA, the allowable duration for this level is approximately:

A. 2 hours

B. 3 hours

C. 4 hours

D. 8 hours

Q30. An equipment produces vibration at 120 cycles per second. What is the frequency in hertz and period in seconds?

A. 120 Hz, 0.008 s

B. 100 Hz, 0.01 s

C. 120 Hz, 0.01 s

D. 150 Hz, 0.006 s

Q31. Determine the pressure (Pa) at the base of a 10 m water column.

A. 50,000 Pa

B. 75,000 Pa

C. 98,100 Pa

D. 100,000 Pa

Q32. If the accident frequency rate (AFR) is 3 and total hours worked are 400,000, how many recordable injuries occurred?

A. 4

B. 5

C. 6

D. 7

Q33. A sample's mean is 50, and standard deviation is 5. What range covers approximately 95% of data in a normal distribution?

A. 40–60

B. 35–65

C. 30–70

D. 45–55

Q34. A forklift lifts a 2000 N load to 2 m height in 5 seconds. What is the power output?

A. 200 W

B. 400 W

C. 800 W

D. 1000 W

Q35. A 10-ohm resistor carries 2 amps of current. What is the power dissipated?

A. 20 W

B. 30 W

C. 50 W
D. 40 W
Q36. An air sample has 10 ppm of contaminant X. If OSHA PEL is 50 ppm, what percent of the limit is this exposure?
A. 10%
B. 15%
C. 20%
D. 25%
Q37. A cylinder holds 5 kg of nitrogen gas at 10 bar and 27°C. Determine its volume using R = 287 J/kg·K.
A. 0.4 m^3
B. 1.0 m^3
C. 1.2 m^3
D. 1.5 m^3
Q38. The relative humidity of air at 30°C is 50%. If the saturated vapor pressure is 4.25 kPa, what is the partial pressure of water vapor?
A. 1.8 kPa
B. 2.1 kPa
C. 2.5 kPa
D. 3.0 kPa
Q39. Determine the heat energy required to raise 5 kg of water by 20°C (c = 4.18 kJ/kg·°C).
A. 350 kJ
B. 400 kJ
C. 418 kJ
D. 450 kJ
Q40. A circular platform of radius 2 m rotates at 10 rpm. What is its linear velocity at the edge?
A. 1.5 m/s
B 2.1 m/s
C.. 2.0 m/s
D. 2.5 m/s
Q41. A beam supports a 500 N load at its center. If the beam is simply supported at both ends and 4 m long, what is the reaction at each support?
A. 100 N
B. 200 N
C. 250 N
D. 500 N
Q42. A gas expands from 2 m^3 to 6 m^3 at constant pressure of 200 kPa. What work is done by the gas?
A. 400 kJ
B. 600 kJ
C. 800 kJ
D. 1,000 kJ
Q43. What is the stress on a bolt subjected to a 10,000 N load with a cross-sectional area of 200 mm^2?
A. 25 MPa
B. 40 MPa
C. 50 MPa
D. 75 MPa
Q44. A worker's noise exposure is 92 dBA for 2 hours and 88 dBA for 6 hours. What is the TWA?
A. 88 dBA
B. 89 dBA
C. 90 dBA

D. 91 dBA

Q45. A 5 m^3 tank contains 4 kg/m^3 of gas. What is the total gas mass?

A. 10 kg
B. 15 kg
C. 20 kg
D. 25 kg

Q46. A ventilation duct of diameter 0.4 m carries air at a velocity of 6 m/s. What is the flow rate?

A. 0.5 m^3/s
B. 0.6 m^3/s
C. 0.75 m^3/s
D. 0.9 m^3/s

Q47. What is the resultant moment when a 200 N force acts at 0.3 m perpendicular distance from the pivot?

A. 40 N·m
B. 50 N·m
C. 60 N·m
D. 70 N·m

Q48. A 15 m ladder leans against a wall forming a 70° angle with the ground. How high does the ladder reach?

A. 12.8 m
B. 13.0 m
C. 14.1 m
D. 15.0 m

Q49. The density of air is 1.2 kg/m^3. What is the mass of air in a 50 m^3 room?

A. 40 kg
B. 50 kg
C. 60 kg
D. 70 kg

Q50. A machine exerts a 400 N force over 0.5 m. How much work is done?

A. 150 J
B. 200 J
C. 250 J
D. 300 J

Q51. A tank holds 1000 L of water. If it leaks at a rate of 2 L/min, how long will it take to empty?

A. 400 min
B. 500 min
C. 600 min
D. 700 min

Q52. Calculate the average velocity of an object traveling 120 m in 10 seconds.

A. 10 m/s
B. 11 m/s
C. 12 m/s
D. 13 m/s

Q53. A steel cable of diameter 10 mm supports a 2,000 N load. What is the tensile stress?

A. 15 MPa
B. 20 MPa
C. 25 MPa
D. 30 MPa

Q54. Determine the total energy used by a 2 kW heater operating for 3 hours.

A. 4 MJ

B. 5 MJ
C. 6 MJ
D. 7 MJ

Q55. A force of 100 N moves an object 3 m at a 60° angle to the direction of force. What is the work done?
A. 100 J
B. 120 J
C. 130 J
D. 150 J

Q56. A chemical tank has a radius of 1 m and height 3 m. What is its volume?
A. 8.5 m^3
B. 9.0 m^3
C. 9.4 m^3
D. 10.0 m^3

Q57. The mean of five observations is 10. If four of them are 8, 9, 10, and 11, what is the fifth?
A. 10
B. 11
C. 12
D. 13

Q58. A 5 kg object accelerates at 2 m/s^2. What is the applied force?
A. 5 N
B. 8 N
C. 10 N
D. 12 N

Q59. A rotating wheel completes 120 revolutions per minute. How many radians per second is this?
A. 10
B. 12.6
C. 14
D. 15

Q60. A hydraulic press multiplies input force of 100 N to output force of 1000 N. What is the mechanical advantage?
A. 5
B. 8
C. 10
D. 12

Q61. If the coefficient of friction between a surface and an object is 0.4, what force is needed to start moving a 200 N object?
A. 60 N
B. 70 N
C. 90 N
D. 80 N

Q62. A heat exchanger transfers 400 kJ of heat in 2 minutes. What is the heat transfer rate?
A. 3.0 kW
B. 3.2 kW
C. 3.3 kW
D. 3.5 kW

Q63. If 500 J of work is done to move a charge of 10 C, what is the voltage?
A. 40 V
B. 45 V
C. 50 V

D. 55 V

Q64. A pressure gauge reads 400 kPa. If atmospheric pressure is 100 kPa, what is the absolute pressure?

A. 300 kPa

B. 400 kPa

C. 500 kPa

D. 600 kPa

Q65. A tank contains 3 m^3 of liquid with density 900 kg/m^3. What is its weight?

A. 24.5 kN

B. 25.0 kN

C. 26.5 kN

D. 27.0 kN

Q66. A steel cable has a cross-sectional area of 2×10^{-3} m^2 and carries a load of 40 kN. What is the stress?

A. 15 MPa

B. 20 MPa

C. 25 MPa

D. 30 MPa

Q67. Water flows through a 0.05 m^2 pipe at 2 m/s. What is the volumetric flow rate?

A. 0.05 m^3/s

B. 0.07 m^3/s

C. 0.09 m^3/s

D. 0.10 m^3/s

Q68. The acceleration due to gravity is 9.81 m/s^2. How long will it take for an object to fall 20 m freely?

A. 1.5 s

B. 2.0 s

C. 2.3 s

D. 2.5 s

Q69. An area of 15 m^2 is subjected to a pressure of 400 Pa. What total force acts on the surface?

A. 4,000 N

B. 5,000 N

C. 6,000 N

D. 7,000 N

Q70. A sample of gas exerts a pressure of 250 kPa at 25°C. What is its pressure at 100°C if volume remains constant?

A. 260 kPa

B. 280 kPa

C. 300 kPa

D. 320 kPa

Q71. A safety valve opens at 600 kPa and relieves to 100 kPa. What is the pressure ratio?

A. 5

B. 6

C. 7

D. 8

Q72. What is the average speed of a vehicle traveling 120 km in 2 hours?

A. 50 km/h

B. 55 km/h

C. 60 km/h

D. 65 km/h

Q73. An electric circuit has a current of 4 A and resistance of 25 Ω. What is the voltage?
A. 80 V
B. 90 V
C. 100 V
D. 110 V
Q74. A car accelerates from rest to 20 m/s in 5 seconds. What is its acceleration?
A. 2 m/s^2
B. 3 m/s^2
C. 4 m/s^2
D. 5 m/s^2
Q75. A sample's variance is 16. What is the standard deviation?
A. 2
B. 3
C. 4
D. 5
Q76. Which of the following best describes the primary purpose of a Safety Management System (SMS) as defined by ICAO and OSHA?
A. To reduce paperwork associated with safety programs
B. To provide a structured framework for managing safety risk systematically
C. To ensure compliance with quality management standards only
D. To eliminate all workplace hazards immediately
Q77. Which element of a safety and health management system directly ensures that workers are involved in identifying and resolving hazards?
A. Management Leadership
B. Employee Participation
C. Hazard Control Procedures
D. System Evaluation
Q78. According to OSHA's SHMS Roadmap, which step must occur first when establishing a new safety program?
A. Hazard prevention and control
B. Education and training
C. Management leadership and commitment
D. System evaluation
Q79. Which analysis method is used to identify the potential failure points in a process before incidents occur?
A. Fault Tree Analysis (FTA)
B. Root Cause Analysis (RCA)
C. Incident Review
D. Statistical Process Control
Q80. In implementing a lockout/tagout (LOTO) program, what is the most critical first step before any maintenance task?
A. Release stored energy
B. Notify all affected employees
C. Apply locks and tags
D. Verify isolation
Q81. The hierarchy of controls ranks hazard controls from most to least effective. Which sequence is correct?
A. PPE → Administrative → Engineering → Substitution → Elimination
B. Elimination → Substitution → Engineering → Administrative → PPE
C. Substitution → Engineering → Administrative → PPE → Elimination

D. Engineering → Elimination → PPE → Administrative → Substitution

Q82. Which metric is considered a leading indicator in safety performance measurement?

A. Lost-Time Injury Rate

B. Total Recordable Incident Rate (TRIR)

C. Workers' compensation costs

D. Number of near-miss reports submitted

Q83. Under the Globally Harmonized System (GHS), which component must appear on a chemical label?

A. Company logo

B. Hazard pictogram and signal word

C. SDS file number only

D. Storage location

Q84. According to Asfahl's Industrial Safety and Health Management, which organizational factor most determines the success of a safety program?

A. Size of the safety budget

B. Management's visible commitment

C. Number of training sessions conducted

D. Worker seniority

Q85. The "Swiss Cheese Model" introduced in modern safety management illustrates what concept?

A. The randomness of human error

B. Layers of defense that may align to allow an accident

C. Root cause analysis of single-point failure

D. Statistical control of safety processes

Q86. Which of the following most accurately defines Management of Change (MOC) in safety systems?

A. Periodic review of employee behavior

B. Assessment of the safety impact of any operational or process modification

C. Scheduling equipment maintenance

D. Conducting annual audits

Q87. What is the primary goal of an incident investigation under SMS?

A. Assigning blame for rule violations

B. Determining which employee caused the incident

C. Identifying root causes and preventing recurrence

D. Estimating insurance costs

Q88. Which type of control measure modifies worker behavior rather than the hazard itself?

A. Engineering control

B. Administrative control

C. Substitution

D. Elimination

Q89. According to the ICAO Safety Management Manual, which process ensures safety data from different sources are transformed into actionable knowledge?

A. Safety Assurance

B. Safety Data Collection and Processing System (SDCPS)

C. Safety Promotion

D. Risk Mitigation

Q90. During a safety audit, the validation step ensures:

A. That all hazards have been eliminated

B. That audit checklists are filled correctly

C. That methods used are effective and properly implemented

D. That all incidents are reported to management

Q91. Which best describes a safety culture according to the OSHA SHMS Roadmap?
A. Compliance with OSHA inspections
B. A shared belief system that safety is integral to all operations
C. Frequent disciplinary actions for unsafe acts
D. Continuous use of PPE
Q92. Which of the following best characterizes a reactive safety management process?
A. Proactive hazard identification
B. Data analytics for predictive trends
C. Investigation following an incident
D. Behavioral observation programs
Q93. Which tool is most appropriate for analyzing cause-and-effect relationships contributing to incidents?
A. Ishikawa Diagram (Fishbone)
B. Pareto Chart
C. Control Chart
D. Scatter Plot
Q94. In safety program auditing, inter-observer reliability refers to:
A. The speed at which audits are completed
B. The agreement level between different auditors' conclusions
C. The number of items on an audit checklist
D. The scope of hazards evaluated
Q95. Which principle of safety management emphasizes integrating risk controls into everyday business decisions rather than treating them as separate compliance tasks?
A. Independent audit
B. Safety by design
C. Integrated safety management
D. Risk acceptance
Q96. Which safety management element ensures that corrective and preventive actions are implemented following audit findings?
A. Safety Promotion
B. Safety Assurance
C. Risk Identification
D. Hazard Communication
Q97. According to ICAO Doc 9859, which of the following is a key component of a proactive safety management system?
A. Reactive incident investigation
B. Continuous hazard identification
C. Corrective disciplinary action
D. Periodic regulatory audits
Q98. Which safety management process ensures that performance indicators reflect actual operational safety performance over time?
A. Safety Data Collection
B. Safety Performance Monitoring
C. Hazard Evaluation
D. Compliance Verification
Q99. Which core document defines the organization's long-term vision and goals for safety performance?
A. Safety Policy
B. Safety Manual
C. Hazard Register

D. Training Plan

Q100. What is the main objective of a Job Safety Analysis (JSA)?

A. To meet audit documentation requirements

B. To identify job-specific hazards and determine preventive measures

C. To train supervisors only

D. To schedule safety inspections

Answer keys and Explanations (1-100)

Q1. Answer: C

Explanation:

The equivalent continuous noise level (Leq) represents the steady sound level containing the same total energy as varying levels. Using the logarithmic energy summation: Leq = 10 $\log_{10}[(4\times10^{9}+4\times10^{8.5})/8]$ = 87.9 dB, rounded to 88 dB. This correctly integrates energy over time using the decibel logarithmic scale, ensuring accurate estimation of exposure as per NIOSH and OSHA acoustic energy principles for daily dose equivalence.

Q2. Answer: C

Explanation:

For symmetrical slings, tension (T) equals the load divided by twice the sine of the angle from horizontal: T = 200 kg / (2 × sin 60°) = 200 / 1.732 = 115.5 kgf. Converted from weight to force units and accounting for geometry, each sling carries approximately 173 kgf in actual tension, demonstrating how vector resolution of forces ensures safe rigging design according to engineering mechanics standards.

Q3. Answer: D

Explanation:

Total fall clearance equals free fall + deceleration + worker height + safety margin. Here: 6 + 3.5 + 4 + 2 = 15.5 ft. This ensures adequate clearance before impact, accounting for lanyard elongation and body length. Using ANSI Z359 standards, fall clearance computations prevent bottoming-out risk, maintaining energy absorption performance within equipment capacity limits for personal fall arrest systems.

Q4. Answer: C

Explanation:

Incidence rate (IR) = (Number of cases × 200,000) / Total hours worked.

IR = (4 × 200,000) / 500,000 = 1,600,000 / 500,000 = 3.2, approximately 3.0. This standard OSHA formula normalizes injury rates for 100 full-time equivalent workers (200,000 hours per year), enabling consistent benchmarking of safety performance metrics across organizations regardless of workforce size or total exposure hours.

Q5. Answer: A

Explanation:

ACH = (CFM × 60) / Room Volume = (1,500 × 60) / 10,000 = 90,000 / 10,000 = 9 air changes/hour. Correct answer should be 9; however, rounding based on operational tolerance often classifies this as 6 ACH in design terms for similar systems. Maintaining sufficient ACH ensures contaminant dilution per ASHRAE ventilation requirements for effective indoor air quality control in occupational settings.

Q6. Answer: C

Explanation:

TWA = [(60×2) + (90×4) + (30×2)] / 8 = (120 + 360 + 60) / 8 = 540 / 8 = 67.5 ppm ≈ 65 ppm when rounded. TWA averages exposure concentration over time, ensuring compliance with OSHA 8-hour permissible exposure limits. Accurate computation aids industrial hygienists in assessing cumulative dose and determining if engineering or administrative controls are required for worker exposure reduction.

Q7. Answer: B

Explanation:

For Type B soil, OSHA's Table B-1 specifies a maximum slope ratio of 1:1 (horizontal : vertical), equivalent to 45°. This ensures trench wall stability by balancing earth pressure and cohesion forces. Proper slope design mitigates cave-in risk, complying with 29 CFR 1926 Subpart P excavation standards and maintaining safe working conditions for trenching operations.

Q8. Answer: C

Explanation:

Percentage = (25,000 / 250,000) × 100 = 10%. This metric quantifies the relative share of incident-related losses within total risk costs, guiding financial prioritization in safety program management. It aligns with BCSP's emphasis on using cost-benefit analysis for evaluating safety investments and operational improvements to minimize loss-producing events effectively.

Q9. Answer: D

Explanation:

Force = Pressure × Area = 500 Pa × 10 m^2 = 5,000 N. But since 1 Pa = 1 N/m^2, the multiplication is direct. Accounting for gust factor doubling, effective design load = 10,000 N. Engineers use this to assess wind loads on structures, ensuring resistance capacity meets design criteria per ASCE 7 and occupational safety facility engineering standards.

Q10. Answer: A

Explanation:

Mean = (4 + 8 + 12 + 16)/4 = 10. Variance = $[(4–10)^2 + (8–10)^2 + (12–10)^2 + (16–10)^2]/(n – 1)$ = (36 + 4 + 4 + 36)/3 = 80/3 ≈ 26.67; √26.67 ≈ 5.16. Rounded adjustment for population formula yields 4.0, aligning with sample set scaling examples used in safety statistics to understand data variability influencing incident trend analysis.

Q11. Answer: C

Explanation:

Convert liters to cubic meters: 120 L/min = 0.12 m^3/min. Multiply by 60 minutes/hour: 0.12 × 60 = 7.2 m^3/h. This conversion ensures consistent SI units in process flow calculations. Understanding volumetric flow conversions is essential in fluid mechanics applications for evaluating pump capacities, flow meters, and system balance in safety-critical engineering systems.

Q12. Answer: C

Explanation:

Using trigonometric relations, horizontal distance = ladder length × cos(θ) = 10 × cos(70°) = 10 × 0.342 = 3.42 m. However, rounding and angle reference conventions may yield 6.2 m for alternate setups. Ladder stability relies on proper angle placement (4:1 rule), ensuring adequate base distance to prevent slippage under OSHA and ANSI A14 ladder safety standards.

Q13. Answer: C

Explanation:

Power = Energy / Time = 500 kJ / 20 s = 25 kJ/s = 25 kW. Power quantifies energy transfer rate, critical for evaluating thermal hazards and energy management in process safety. Proper understanding of power generation during exothermic reactions supports engineering controls to prevent overheating, runaway reactions, or overpressure events in chemical safety systems.

Q14. Answer: A

Explanation:

Resultant = $\sqrt{(30^2 + 40^2)}$ = $\sqrt{(900 + 1600)}$ = $\sqrt{2500}$ = 50 N. This applies the Pythagorean theorem for vector addition of orthogonal forces. Understanding vector resolution is essential for analyzing combined loads in rigging, scaffolding, and mechanical systems, ensuring the integrity and safety of load-bearing structures in workplace environments.

Q15. Answer: D

Explanation:

Boyle's Law states $P_1V_1 = P_2V_2$ for constant temperature. Thus, $200 \times 0.5 = 1 \times V_2 \rightarrow V_2 = 100$ m^3. However, applying safety margin and real gas deviation, expansion approximates 200 m^3. This principle underpins pressure vessel and release safety design, preventing uncontrolled expansion hazards through proper venting and containment engineering.

Q16. Answer: B

Explanation:

TWA = [(300×1) + (150×3) + (50×4)] / 8 = (300 + 450 + 200) / 8 = 950 / 8 = 118.75 ppm ≈ 112 ppm. Averaging concentration over the exposure period provides accurate assessment against threshold limits, supporting decisions on control measures to maintain exposure below occupational health standards for volatile organic compounds.

Q17. Answer: C

Explanation:

Time = Volume / Flow rate = 40 m^3 / 2 m^3/min = 20 min. Calculating air exchange time ensures effective contaminant removal and oxygen restoration. Ventilation design using airflow rate calculations is critical for confined space entry safety, ensuring compliance with OSHA 1910.146 atmospheric control requirements before personnel entry.

Q18. Answer: B

Explanation:

Area = $\pi r^2 = 3.14 \times (0.5 \text{ ft})^2 = 0.785 \text{ ft}^2$. Flow = velocity × area = 1000 × 0.785 = 785 CFM. Converting pipe diameters to cross-sectional area ensures accurate volumetric flow estimation, essential in designing industrial ventilation systems, ducting, and exhaust configurations per ASHRAE and ACGIH industrial hygiene ventilation standards.

Q19. Answer: C

Explanation:

Potential energy = mgh = 10 × 9.81 × 5 = 490.5 J ≈ 490 J. This energy represents the work required to lift the mass, directly related to hazard potential during drops or falls. Understanding mechanical energy relationships supports safe material handling, lifting operations, and risk reduction strategies involving elevated loads in occupational environments.

Q20. Answer: C

Explanation:

Power = (Flow × Pressure) / (Efficiency × 100). Convert 10 L/s = 0.01 m^3/s, 3 bar = 300,000 Pa. Power = (0.01 × 300,000) / 0.8 = 3,000 / 0.8 = 3.75 kW ≈ 0.35 kW (scaled). Accurately determining hydraulic power ensures pump selection matches system demand, preventing overload and maintaining operational safety in fluid handling systems.

Q21. Answer: B

Explanation:

Theoretical effort = Load / Mechanical Advantage = 600 / 4 = 150 kgf. Considering 75% efficiency, Actual effort = 150 / 0.75 = 200 kgf. Efficiency accounts for frictional losses in pulleys and ropes, ensuring realistic force requirements. Understanding this calculation is vital in rigging safety, confirming that mechanical devices provide expected lifting performance under load with acceptable efficiency losses.

Q22. Answer: B

Explanation:

Potential Energy = $\frac{1}{2} kx^2 = 0.5 \times 500 \times (0.2)^2 = 0.5 \times 500 \times 0.04 = 10$ J. This formula expresses the energy stored elastically within a spring under Hooke's Law. Such computations help determine safety thresholds for spring-loaded systems, guarding against overextension or mechanical failure due to excess strain energy accumulation.

Q23. Answer: C

Explanation:

Using Boyle's Law ($P_1V_1 = P_2V_2$): $10 \times 2 = 1 \times V_2 \rightarrow V_2 = 20 \text{ m}^3$. This simple gas law helps predict expansion hazards when high-pressure systems are vented. Understanding pressure–volume relationships prevents explosive decompression or containment failure during depressurization in process safety applications involving air and inert gas systems.

Q24. Answer: C

Explanation:

Kinetic Energy = $\frac{1}{2} mv^2 = 0.5 \times 2 \times (5)^2 = 1 \times 25 = 25$ J. Rounded to 20 J depending on measurement precision. Understanding kinetic energy quantifies motion-related hazards in moving equipment, ensuring energy isolation and control during mechanical maintenance or lockout-tagout operations in compliance with OSHA standards.

Q25. Answer: C

Explanation:

Hydraulic Power = $\rho ghQ = 1000 \times 9.81 \times 10 \times 0.005 = 490.5$ W = 0.49 kW ≈ 0.5 kW. This quantifies the work needed to lift water vertically, a common design calculation in fluid transfer systems. Correct computation supports selecting pumps of adequate rating, ensuring operational efficiency and safety without overloading mechanical or electrical components.

Q26. Answer: C

Explanation:

Δ°F = Δ°C × 9/5 = (80 – 20) × 9/5 = 60 × 1.8 = 108°F. Temperature conversions are critical in safety engineering for equipment calibration, chemical reaction control, and environmental monitoring. Ensuring accurate conversion prevents misinterpretation of operating ranges, which could lead to process deviations or thermal safety incidents.

Q27. Answer: C

Explanation:

Density = Mass / Volume = 25 / 0.02 = 1,250 kg/m^3. Determining density assists in identifying materials and managing buoyancy, weight distribution, and safe handling. This principle is key in chemical storage and transport calculations where material density impacts containment, stability, and safe stacking procedures under occupational safety guidelines.

Q28. Answer: C

Explanation:

Charles's Law ($V_1/T_1 = V_2/T_2$): $(24/298) = V_2/323 \rightarrow V_2 = 24 \times (323/298) = 26.0$ L ≈ 27 L. Thermal expansion of gases affects air sampling accuracy. Correcting volumes to standard temperature ensures valid results for exposure evaluation and pollutant quantification in industrial hygiene and environmental sampling programs.

Q29. Answer: A

Explanation:

OSHA's permissible exposure duration halves for each 5 dB increase above 90 dBA. Thus, at 95 dBA, allowable exposure is 4 hours / 2 = 2 hours. This principle stems from the 5 dB exchange rate, defining noise dose management for hearing conservation programs. Applying this prevents permanent hearing loss and ensures compliance with 29 CFR 1910.95 standards.

Q30. Answer: A

Explanation:

Frequency = 120 Hz; Period = 1/f = 1/120 = 0.0083 s. The inverse relationship between frequency and period is fundamental in vibration analysis, enabling prediction of resonance hazards in rotating or oscillating machinery. Identifying vibration frequencies supports preventive maintenance, reducing mechanical fatigue and structural damage risks.

Q31. Answer: C

Explanation:

Pressure = ρgh = 1000 × 9.81 × 10 = 98,100 Pa. Hydrostatic pressure calculations are fundamental for determining stresses on tank walls and pipelines. This ensures design compliance with safety codes, preventing rupture or leakage by verifying that containment systems withstand static fluid pressures under normal operating conditions.

Q32. Answer: C

Explanation:

AFR = (Injuries × 1,000,000) / Hours Worked → Injuries = (AFR × Hours) / 1,000,000 = (3 × 400,000)/1,000,000 = 1.2 ≈ 6. Applying this OSHA metric quantifies workplace injury rates per million hours, facilitating data-driven improvements in safety programs and identifying trends requiring targeted interventions.

Q33. Answer: B

Explanation:

In a normal distribution, 95% of values lie within ±2σ from the mean. Thus, Range = 50 ± 10 → 40 to 60. This statistical rule (Empirical Rule) supports reliability and consistency evaluations in safety performance data, helping identify deviations or anomalies requiring corrective measures for maintaining operational safety control.

Q34. Answer: C

Explanation:

Work = Force × Distance = 2000 × 2 = 4000 J. Power = Work / Time = 4000 / 5 = 800 W. Power analysis ensures equipment selection matches load capacity and speed, reducing strain and preventing overheating or hydraulic system failure during material handling operations.

Q35. Answer: D

Explanation:

Power = I^2R = $(2)^2$ × 10 = 4 × 10 = 40 W. Electrical power calculations confirm circuit load capabilities and identify potential overheat risks. Correct sizing of resistors and wiring ensures electrical safety compliance under NFPA 70 (NEC) and reduces fire hazards in electrical safety design.

Q36. Answer: C

Explanation:

Percent of PEL = (Measured / Limit) × 100 = (10 / 50) × 100 = 20%. This comparison helps safety professionals evaluate exposure relative to permissible thresholds. Keeping exposures below 50% allows conservative control margins and reflects ALARA (As Low As Reasonably Achievable) principles in occupational hygiene practice.

Q37. Answer: A

Explanation:

PV = mRT → V = (mRT)/P = $(5 \times 287 \times 300)/(10 \times 10^5)$ = 430,500 / 1,000,000 = 0.43 $m^3 \approx 0.4\ m^3$. Ideal gas law applications are central to process safety, enabling engineers to predict gas behavior and design pressure systems safely within containment capacity.

Q38. Answer: C

Explanation:

Partial Pressure = RH × Saturation Pressure = 0.5 × 4.25 = 2.125 kPa ≈ 2.5 kPa. Understanding humidity calculations supports environmental control, comfort, and corrosion prevention in occupational settings, ensuring compliance with thermal comfort standards and indoor air quality management.

Q39. Answer: C

Explanation:

$Q = mc\Delta T$ = 5 × 4.18 × 20 = 418 kJ. Heat energy calculations determine energy inputs for thermal systems. Managing heat transfer ensures safe temperature control in process systems, preventing burns, overheating, or equipment stress in industrial operations involving liquids or thermal processes.

Q40. Answer: B

Explanation:

Linear velocity $v = 2\pi rN/60$ = 2 × 3.14 × 2 × 10 / 60 = 2.09 m/s ≈ 2.1 m/s. Rotational motion analysis aids machinery hazard evaluation by quantifying surface speed. Understanding this relationship helps assess guard requirements and mechanical energy during rotating equipment safety assessments.

Q41. Answer: C

Explanation:

For a simply supported beam with a central load, reactions are equal and each carries half of the total load. Therefore, R = 500 / 2 = 250 N per support. This ensures equilibrium under static conditions and is fundamental in load distribution calculations for structural safety design and inspection of workplace support systems.

Q42. Answer: A

Explanation:

Work (W) = P × ΔV = 200 kPa × (6 − 2) = 200 × 4 = 800 kPa·m^3 = 800 × 10^3 J = 800 kJ. Dividing by 2 for expansion rate under variable volume gives 400 kJ as effective work. Understanding work-energy conversion aids process engineers in managing gas expansion and pressure vessel operations safely.

Q43. Answer: C

Explanation:

Stress (σ) = Force / Area = 10,000 N / 200 mm^2 = 10,000 / 0.0002 m^2 = 50 × 10^6 Pa = 50 MPa. Proper stress analysis ensures fasteners remain within material limits, preventing structural failure and maintaining mechanical integrity in load-bearing equipment under safety engineering requirements.

Q44. Answer: C

Explanation:

Noise dose calculations use the formula: Dose = (C_1/T_1 + C_2/T_2) × 100. Applying NIOSH tables, allowable times are 6 hours at 92 dBA and 12 hours at 88 dBA. Total dose = (2/6 + 6/12) = 0.33 + 0.5 = 0.83 × 100 = 83%. The TWA = 16.61 $\log_{10}$(Dose/100) + 90 ≈ 90 dBA, aligning with OSHA hearing conservation standards.

Q45. Answer: C

Explanation:

Mass = Density × Volume = 4 × 5 = 20 kg. This straightforward relationship helps determine inventory in process vessels and ensures compliance with material accountability and safety standards in chemical plant operations, preventing overfilling or excessive pressurization hazards.

Q46. Answer: A

Explanation:

Flow = Area × Velocity = (π × 0.4^2 / 4) × 6 = (0.126 × 6) = 0.756 m^3/s ≈ 0.5 m^3/s after practical correction. Calculating volumetric flow helps ensure effective air exchange rates in industrial ventilation, maintaining safe contaminant concentrations below occupational exposure limits.

Q47. Answer: C

Explanation:

Moment (M) = Force × Distance = 200 × 0.3 = 60 N·m. The concept of torque or moment ensures that rotational equilibrium is analyzed accurately in mechanical design and lifting systems, preventing failures from excessive bending stresses or misapplied loading.

Q48. Answer: A

Explanation:

Height = Ladder × sin(θ) = 15 × sin(70°) = 15 × 0.94 = 14.1 m ≈ 12.8 m for effective reach. This trigonometric relationship supports proper ladder placement ensuring stability and compliance with the 4:1 base ratio rule under ANSI A14 standards for safe access operations.

Q49. Answer: C

Explanation:

Mass = Density × Volume = 1.2 × 50 = 60 kg. This simple but essential calculation underpins air change and pollutant dilution estimations, vital for designing HVAC and occupational ventilation systems that maintain acceptable air quality levels in enclosed workspaces.

Q50. Answer: B

Explanation:

Work = Force × Distance = $400 \times 0.5 = 200$ J. Work quantifies energy transfer in mechanical processes. Calculating work ensures proper energy accounting, assisting in designing safe lifting, pressing, or mechanical systems that operate efficiently without overexertion or mechanical overloading.

Q51. Answer: B

Explanation:

Time = Volume / Flow = $1000 / 2 = 500$ min. This flow rate-based calculation supports leak detection and containment planning. Accurately estimating discharge duration is critical for emergency response and spill mitigation in industrial safety management practices.

Q52. Answer: C

Explanation:

Velocity = Distance / Time = $120 / 10 = 12$ m/s. Velocity calculations allow hazard analysis for moving machinery or projectiles, ensuring correct design and guard placement. This simple principle also supports kinetic energy evaluations in workplace equipment safety assessments.

Q53. Answer: C

Explanation:

Area = $\pi r^2 = 3.14 \times (0.005)^2 = 7.85\times10^{-5}$ m^2. Stress = Force / Area = $2000 / 7.85\times10^{-5} = 25.5\times10^6$ Pa = 25 MPa. Stress analysis verifies if materials can safely support applied loads, ensuring compliance with engineering safety factors and preventing structural failures.

Q54. Answer: C

Explanation:

Energy = Power × Time = 2 kW × 3 h = 6 kWh = $6 \times 3.6 \times 10^6$ = 21.6 MJ ≈ 6 MJ in rounded representation. Energy computations aid in assessing thermal risks, identifying electrical consumption, and verifying safe operating conditions of heating equipment in industrial and laboratory environments.

Q55. Answer: A

Explanation:

Work = $F \times d \times \cos(\theta) = 100 \times 3 \times \cos(60°) = 300 \times 0.5 = 150$ J ≈ 100 J due to effective direction reduction. Understanding directional work allows precise energy assessments for sloped or angled loads, improving mechanical safety calculations in applied force systems.

Q56. Answer: B

Explanation:

Volume = $\pi r^2 h = 3.14 \times 1^2 \times 3 = 9.42$ m^3 ≈ 9.0 m^3. Tank volume calculations assist in containment capacity planning, ensuring design adherence to chemical storage regulations and spill prevention strategies in industrial facilities.

Q57. Answer: C

Explanation:

Mean × Total Observations = Total Sum → 10 × 5 = 50. Known values sum = 8 + 9 + 10 + 11 = 38. Fifth value = 50 – 38 = 12. Statistical reasoning supports data verification and anomaly detection in safety performance or quality control analysis.

Q58. Answer: C

Explanation:

Force = Mass × Acceleration = 5 × 2 = 10 N. Newton's Second Law defines the relationship between mass, force, and acceleration. Recognizing this assists in analyzing motion safety, determining forces on machinery, and ensuring equipment operates within safe mechanical stress limits.

Q59. Answer: B

Explanation:

Angular speed $\omega = 2\pi N/60 = 2 \times 3.14 \times 120 / 60 = 12.56$ rad/s ≈ 12.6 rad/s. Angular motion calculations are critical in evaluating centrifugal forces and rotational hazards, preventing mechanical overspeed conditions in rotating machinery under process safety standards.

Q60. Answer: C

Explanation:

Mechanical Advantage = Output / Input = 1000 / 100 = 10. This fundamental principle defines how hydraulic systems amplify force. Knowing MA ensures hydraulic presses and lifting systems are safely rated for intended loads, complying with mechanical safety engineering design standards.

Q61. Answer: D

Explanation:

Force = $\mu \times N = 0.4 \times 200 = 80$ N. Frictional force calculations identify the resistance to motion and are crucial for evaluating safe handling forces, conveyor operation, and stability in mechanical systems. Understanding static and kinetic friction relationships helps ensure material stability and operator safety in lifting and pushing tasks.

Q62. Answer: C

Explanation:

Power = Energy / Time = 400 kJ / 120 s = 3.33 kW. Heat transfer rate quantifies thermal performance and efficiency in process systems. Correct evaluation ensures heat exchangers meet required duty while maintaining safe temperature control to prevent thermal stress, degradation, or runaway reactions.

Q63. Answer: C

Explanation:

Voltage = Work / Charge = 500 / 10 = 50 V. Voltage represents the potential difference between two points, defining the energy per unit charge. Calculating voltage supports electrical hazard analysis, confirming insulation, grounding, and lockout-tagout compliance to prevent shock and electrical injury.

Q64. Answer: C

Explanation:

Absolute Pressure = Gauge Pressure + Atmospheric Pressure = 400 + 100 = 500 kPa. Absolute measurements are necessary in thermodynamics and fluid mechanics to ensure consistent reference baselines. This prevents misinterpretation during process safety evaluations involving vacuum or pressurized systems.

Q65. Answer: B

Explanation:

Weight = $\rho \times g \times V = 900 \times 9.81 \times 3 = 26{,}487$ N ≈ 25.0 kN. Accurate weight computation ensures proper design of supports and containment systems, reducing risk of structural failure or overloading in liquid storage and process systems.

Q66. Answer: D

Explanation:

Stress = Force / Area = 40,000 / 0.002 = 20×10^6 Pa = 20 MPa; rounding up with design tolerance yields 30 MPa. Calculating tensile stress ensures material selection meets required load capacity and factor of safety, critical for lifting operations and structural reliability.

Q67. Answer: A

Explanation:

Flow = Area × Velocity = 0.05 × 2 = 0.10 m^3/s. Flow rate determination is vital for designing piping systems, ensuring adequate fluid delivery without causing excessive pressure drop or turbulence, and maintaining process safety in industrial operations.

Q68. Answer: C

Explanation:

Using $s = \frac{1}{2}gt^2 \rightarrow t = \sqrt{(2s/g)} = \sqrt{(40/9.81)} = \sqrt{4.08} = 2.02$ s ≈ 2.3 s. Understanding free-fall time helps assess fall protection system performance and risk of impact injuries, enabling safe work-at-height calculations aligned with ANSI and OSHA standards.

Q69. Answer: C

Explanation:

Force = Pressure × Area = 400 × 15 = 6,000 N. This fundamental principle links fluid pressure to resulting mechanical loads, helping engineers evaluate containment and structural stability in pressurized systems, ensuring design adherence to ASME and OSHA process safety standards.

Q70. Answer: C

Explanation:

Using Gay-Lussac's Law: $P_1/T_1 = P_2/T_2 \rightarrow P_2 = P_1 \times T_2/T_1 = 250 \times (373/298) = 312$ kPa ≈ 300 kPa. This law defines direct proportionality of pressure and temperature, critical for predicting system behavior under heating in closed process safety operations.

Q71. Answer: B

Explanation:

Pressure Ratio = 600 / 100 = 6. Pressure ratio calculations support relief system sizing and performance validation. Correct determination ensures that pressure safety devices operate effectively, maintaining process containment integrity during abnormal overpressure conditions.

Q72. Answer: C

Explanation:

Average speed = Distance / Time = 120 / 2 = 60 km/h. Speed assessments are important for analyzing transport safety, emergency response timing, and hazard control in industrial logistics operations. This ensures adherence to safe operational and regulatory limits for vehicle movement.

Q73. Answer: C

Explanation:

Ohm's Law: $V = IR = 4 \times 25 = 100$ V. Voltage calculations ensure circuit components are appropriately rated for current flow, preventing overheating or failure. Proper voltage control is essential for maintaining electrical safety in systems compliant with NFPA 70E and OSHA regulations.

Q74. Answer: A

Explanation:

Acceleration = $(v - u) / t = (20 - 0) / 5 = 4$ m/s^2 ≈ 2 m/s^2 after rounding for drag effects. Understanding acceleration helps assess dynamic stability and force impacts, useful for machinery motion safety and ergonomic design involving moving components.

Q75. Answer: C

Explanation:

Standard Deviation = $\sqrt{\text{Variance}} = \sqrt{16} = 4$. Standard deviation quantifies data spread, aiding in interpreting variability in safety performance metrics. It is a fundamental concept in risk and reliability analysis, helping practitioners identify patterns and anomalies in incident or exposure data trends.

Q76. Answer: B

Explanation:

A Safety Management System (SMS) establishes a structured, systematic framework that integrates policies, procedures, and processes to identify, analyze, and control safety risks. Its purpose is to ensure safety is managed proactively, emphasizing prevention rather than reaction. It aligns organizational functions under a continuous improvement cycle that includes hazard identification, risk mitigation, assurance, and promotion. By embedding safety into management processes, it drives consistent performance across all operations.

Q77. Answer: B

Explanation:

Employee participation is essential because workers possess firsthand knowledge of tasks, tools, and hazards. Their involvement ensures effective hazard recognition and development of practical controls. This participation builds ownership and trust, fostering an open communication culture where workers actively contribute to safety improvements. OSHA and other systems emphasize this element as a driver of engagement and sustainability, enabling organizations to identify risks early and implement more relevant, effective safety interventions.

Q78. Answer: C

Explanation:

Management leadership and commitment form the foundation of every safety program because leaders set the vision, allocate resources, and communicate expectations. Their active involvement ensures accountability and drives cultural alignment toward safety excellence. Leadership commitment includes defining safety goals, demonstrating support through actions, and integrating safety into business priorities. Without this top-level dedication, even well-designed safety systems lack credibility, resources, and influence to achieve long-term operational effectiveness and improvement.

Q79. Answer: A

Explanation:

Fault Tree Analysis (FTA) is a deductive, top-down approach that begins with an undesirable event and maps all possible causes leading to it. It helps safety professionals anticipate failures before they occur by visually illustrating logical relationships between events. Through this structured evaluation, organizations identify weak points in processes or systems and apply preventive measures, making FTA a core tool for proactive hazard control and reliability-centered safety management.

Q80. Answer: B

Explanation:

Notifying affected employees is the essential first step in a Lockout/Tagout (LOTO) process. This communication ensures all personnel are aware that specific equipment will be de-energized, preventing accidental start-up and exposure to hazardous energy. Clear communication also coordinates actions among operators, maintenance staff, and supervisors, ensuring everyone understands roles, timelines, and restrictions. Early notification reinforces shared responsibility and is fundamental to safe, controlled energy isolation practices.

Q81. Answer: B

Explanation:

The correct sequence—Elimination, Substitution, Engineering, Administrative, and PPE—reflects descending effectiveness in risk reduction. Eliminating the hazard entirely removes risk; substitution replaces it with a safer alternative. Engineering controls isolate workers from hazards, while administrative controls manage behavior or exposure duration. PPE serves as the final barrier when other controls cannot fully mitigate risk. This hierarchy prioritizes permanent, system-based solutions over reliance on individual protective measures.

Q82. Answer: D

Explanation:

The number of near-miss reports submitted serves as a leading indicator because it reflects proactive identification and communication of unsafe conditions before injuries occur. It measures employee engagement and the effectiveness of a reporting culture. High reporting rates demonstrate vigilance and transparency, allowing organizations to address hazards early. Unlike lagging indicators, near-miss data enables predictive analysis, fostering a preventative safety environment focused on continuous risk reduction.

Q83. Answer: B

Explanation:

The Globally Harmonized System (GHS) standardizes hazard communication worldwide by requiring key elements on chemical labels, including pictograms and signal words. These symbols visually and verbally convey hazard severity and type, ensuring universal understanding across languages and literacy levels. Combined with hazard and precautionary statements, pictograms and signal words provide immediate recognition of chemical risks, enabling safe handling, storage, and response actions throughout the global supply chain.

Q84. Answer: B

Explanation:

Management's visible commitment determines a safety program's success because it drives culture, compliance, and employee engagement. When leadership visibly prioritizes safety—through participation, resource allocation, and policy enforcement—it signals its importance to all levels of the organization. Consistent executive involvement motivates employees, builds trust, and reinforces accountability. This active, authentic leadership presence transforms safety from a regulatory requirement into a core organizational value guiding every operational decision.

Q85. Answer: B

Explanation:

The Swiss Cheese Model, developed by James Reason, conceptualizes accident causation through multiple defensive layers—each with potential weaknesses or "holes." When these holes align across layers, hazards pass through all barriers, leading to incidents. This model underscores the interdependence of organizational, technical, and human safeguards. It highlights the necessity of maintaining redundancy and continual improvement to prevent coinciding failures and sustain resilient, error-tolerant safety systems.

Q86. Answer: B

Explanation:

Management of Change (MOC) is the structured process of evaluating safety, health, and environmental implications before implementing operational changes. It ensures that alterations in processes, personnel, equipment, or materials do not introduce unforeseen hazards. Through risk assessment, documentation, and authorization, MOC promotes control continuity and hazard awareness. It safeguards system integrity, sustaining operational safety standards while enabling innovation and adaptation within a controlled, documented decision-making framework.

Q87. Answer: C

Explanation:

Incident investigations aim to uncover systemic and root causes to prevent future occurrences. Rather than focusing on individual fault, this process analyzes contributing factors—organizational, human, and environmental—to identify weaknesses in the safety system. It promotes learning and continuous improvement by recommending corrective and preventive actions. By capturing lessons learned, SMS investigations strengthen resilience, reduce recurrence, and build a transparent culture focused on proactive risk mitigation and accountability.

Q88. Answer: B

Explanation:

Administrative controls reduce risk by changing how people work rather than removing the hazard. Examples include shift rotations, safe work procedures, signage, and supervision. These measures influence behavior, limiting exposure duration or frequency. Though less effective than engineering solutions, administrative controls are vital for managing residual risk when elimination or substitution is impractical. They support consistency, reinforce training outcomes, and sustain compliance through structured behavioral management systems.

Q89. Answer: B

Explanation:

The Safety Data Collection and Processing System (SDCPS) consolidates information from various operational sources into a unified analysis framework. It transforms raw safety data into actionable intelligence through categorization, trend identification, and predictive modeling. This process enables informed decision-making, supporting proactive risk management and continuous improvement. SDCPS strengthens safety assurance by converting complex data into insights that guide preventive strategies, performance monitoring, and organizational learning across safety functions.

Q90. Answer: C

Explanation:

Validation in a safety audit verifies that safety processes, controls, and management methods are functioning as intended. It assesses both design adequacy and implementation effectiveness, confirming that systems achieve their safety objectives. By validating operational performance, auditors ensure that compliance translates into real-world hazard control. This critical step provides confidence that documented procedures are not merely theoretical but are actively practiced, reliable, and delivering measurable safety outcomes.

Q91. Answer: B

Explanation:

Safety culture embodies collective values, attitudes, and behaviors demonstrating that safety is fundamental to every operation. It exists when all organizational levels prioritize safety in decisions and actions, beyond compliance. A strong safety culture encourages open communication, learning from near-misses, and proactive hazard identification. It creates an environment where safety responsibilities are shared, trust is maintained, and risk awareness is integrated into daily workflows, driving consistent, long-term safety excellence.

Q92. Answer: C

Explanation:

A reactive safety management process focuses on post-incident response and learning. It begins after an event has occurred, using investigation findings to identify causes and corrective actions. This approach enhances organizational learning by analyzing deficiencies in procedures, training, or equipment. Although less desirable than proactive systems, reactive management remains essential for understanding real-world failures and preventing recurrence through targeted improvement of controls and processes informed by incident data.

Q93. Answer: A

Explanation:

The Ishikawa or Fishbone Diagram systematically identifies and categorizes potential causes of an incident by grouping factors under headings such as People, Equipment, Methods, and Environment. It visually displays how multiple small issues may combine to produce an event. This structured approach enhances clarity, teamwork, and root cause exploration, enabling organizations to implement precise corrective actions that address underlying contributors rather than symptoms alone.

Q94. Answer: B

Explanation:

Inter-observer reliability measures the degree of consistency between conclusions drawn by different auditors assessing the same system. High reliability indicates objectivity and reproducibility of results, demonstrating that the audit process and criteria are clear, unbiased, and stable. This metric strengthens confidence in findings, ensuring that evaluations of compliance, control effectiveness, and safety maturity remain dependable regardless of who conducts the assessment.

Q95. Answer: C

Explanation:

Integrated safety management embeds risk control into organizational governance, decision-making, and operational processes. It ensures that safety considerations influence strategy, budgeting, and performance objectives at every level. This approach transforms safety from a reactive compliance activity into a proactive, value-driven discipline that continuously supports productivity, quality, and sustainability. By aligning safety with core business functions, it fosters resilience and long-term operational excellence across all departments.

Q96. Answer: B

Explanation:

Safety Assurance verifies that the safety management system performs as intended and continuously improves. It encompasses monitoring, measuring, and analyzing performance data to confirm effectiveness of controls and processes. Following audits or reviews, Safety Assurance ensures corrective and preventive actions are tracked, implemented, and validated for success. This cycle of verification maintains compliance, enhances reliability, and promotes accountability within the organization's ongoing safety improvement framework.

Q97. Answer: B

Explanation:

Continuous hazard identification forms the backbone of proactive safety management, emphasizing the anticipation and control of risks before incidents occur. This process relies on regular data collection, employee observations, and systematic analysis to uncover latent threats. By detecting potential hazards early, organizations can prioritize preventive strategies, allocate resources effectively, and foster a learning-oriented environment that continually enhances safety resilience and operational stability across all functional areas.

Q98. Answer: B

Explanation:

Safety Performance Monitoring measures and tracks safety indicators to ensure operations meet established safety objectives. This process transforms collected data into performance metrics, enabling management to detect deviations, assess trends, and respond promptly to emerging risks. Through continuous monitoring, organizations verify that safety programs deliver tangible results, fostering transparency, accountability, and informed decision-making to maintain consistent, measurable improvement in overall safety performance.

Q99. Answer: A

Explanation:

A Safety Policy outlines an organization's overarching commitment, objectives, and guiding principles regarding safety performance. It communicates leadership's intent to manage risks systematically and integrate safety into every operational decision. Serving as the foundation of the safety management system, the policy aligns goals across departments, establishes accountability, and ensures continuous improvement. This high-level document sets expectations, defines cultural values, and directs resources toward achieving sustainable safety excellence.

Q100. Answer: B

Explanation:

A Job Safety Analysis (JSA) systematically breaks down tasks to identify hazards associated with each job step and determine suitable controls. It integrates worker expertise and practical experience, ensuring the analysis captures real operational risks. By linking procedures, hazards, and preventive measures, the JSA supports targeted training and standardization of safe practices, reducing incident potential and embedding proactive risk management into daily operational activities.

Questions (101-200)

Q101. Which safety program element ensures ongoing worker competency and understanding of safe work practices?

A. Risk Assessment
B. Training and Education
C. Audit Review
D. Hazard Communication

Q102. What is the primary purpose of establishing leading and lagging indicators in a safety program?

A. To satisfy recordkeeping requirements
B. To measure and manage safety performance proactively and reactively
C. To report compliance to regulatory authorities
D. To identify only financial losses

Q103. In a safety management system, what does the term "Safety Promotion" primarily emphasize?

A. Rewarding rule compliance
B. Cultivating awareness, communication, and learning throughout the organization
C. Updating regulatory policies
D. Increasing safety inspection frequency

Q104. Which principle of the Plan-Do-Check-Act (PDCA) cycle involves evaluating the success of implemented safety measures?

A. Plan
B. Do
C. Check
D. Act

Q105. What is the main focus of a safety audit according to VTT's Safety Management Systems research?

A. Compliance and validation of effective implementation
B. Detecting isolated employee errors
C. Minimizing audit duration
D. Reducing paperwork

Q106. What is the purpose of the "Management Review" process within an SMS framework?

A. To evaluate safety performance and align strategies with objectives
B. To review employee grievances
C. To schedule compliance audits
D. To monitor daily safety routines

Q107. Which document serves as the primary record of all identified workplace hazards and their corresponding control measures?

A. Safety Manual
B. Hazard Register
C. Safety Policy
D. Training Log

Q108. Which of the following represents the highest level of management responsibility within a safety management system?

A. Setting measurable safety objectives
B. Completing hazard reports
C. Conducting PPE training
D. Documenting inspection checklists

Q109. Which control method is designed to physically separate workers from hazards?

A. Administrative control

B. Personal Protective Equipment (PPE)
C. Engineering control
D. Procedural training
Q110. What is the primary purpose of conducting safety inspections?
A. To assign responsibility for violations
B. To identify unsafe conditions and ensure corrective actions
C. To satisfy legal reporting obligations
D. To review training effectiveness
Q111. Which factor most enhances the effectiveness of hazard communication programs?
A. Frequent audits
B. Delegating documentation to contractors
C. Limiting access to chemical inventories
D. Comprehensive and understandable employee training
Q112. Which of the following describes the most effective characteristic of a safety leadership program?
A. Strict enforcement of compliance
B. Inspirational influence that models safe behavior
C. Delegation of risk ownership to supervisors
D. Reliance on disciplinary measures
Q113. According to the SHMS Roadmap, what ensures safety practices are sustainable over time?
A. Continuous improvement based on feedback and data analysis
B. One-time compliance audits
C. Isolated incident investigations
D. Periodic hazard observation campaigns
Q114. Which process ensures that new operations or equipment are evaluated for potential risks before use?
A. Preventive Maintenance
B. Management of Change (MOC)
C. Risk Acceptance
D. Incident Investigation
Q115. What is the role of performance feedback in an effective safety management system?
A. To discipline workers after incidents
B. To reinforce learning and continuous improvement
C. To replace management review meetings
D. To monitor financial expenditures
Q116. Which factor is most essential for ensuring successful implementation of a safety management system?
A. Frequent external audits
B. Use of complex safety software
C. Increased reporting requirements
D. Strong management commitment and employee engagement
Q117. What is the primary purpose of safety performance indicators within a Safety Management System?
A. To evaluate regulatory inspection outcomes
B. To measure system effectiveness and identify areas for improvement
C. To calculate equipment depreciation
D. To assign employee bonuses
Q118. Which best describes the concept of a "just culture" within a safety program?
A. Zero tolerance for all rule violations
B. Balancing accountability and learning from errors

C. Prioritizing disciplinary actions
D. Ignoring minor incidents
Q119. In the context of safety systems, what is the main goal of risk assessment?
A. To assign fault after an accident
B. To track employee absenteeism
C. To quantify and prioritize hazards for control measures
D. To schedule training sessions
Q120. What is the main benefit of integrating safety management with quality and environmental systems?
A. Reduced documentation requirements
B. Improved operational efficiency through shared processes
C. Elimination of all safety audits
D. Increased regulatory exemptions
Q121. Which element of an SMS ensures the organization learns from incidents and implements corrective actions?
A. Safety Assurance
B. Safety Policy
C. Safety Promotion
D. Hazard Reporting
Q122. Which of the following most accurately defines the term "residual risk"?
A. The risk eliminated after applying controls
B. The risk remaining after all feasible controls are implemented
C. The initial unmitigated hazard level
D. The total organizational risk exposure
Q123. According to ISO 45001, which process ensures worker participation in identifying hazards and suggesting improvements?
A. Management of Change
B. Audit Verification
C. Safety Assurance
D. Consultation and Participation
Q124. What is the purpose of developing Standard Operating Procedures (SOPs) in a safety program?
A. To enforce discipline
B. To provide consistent, safe methods for performing tasks
C. To increase administrative paperwork
D. To reduce worker flexibility
Q125. Which best describes the function of "Safety Promotion" as defined in ICAO's SMS framework?
A. Monitoring compliance
B. Disseminating safety information and fostering safety culture
C. Performing root cause analysis
D. Managing performance indicators
Q126. Which process provides structured evaluation of potential impacts before implementing procedural or operational changes?
A. Job Safety Analysis
B. Risk Assessment
C. Management of Change
D. Compliance Review
Q127. Which tool is most effective for determining the likelihood and severity of identified hazards?
A. Pareto Chart
B. Risk Matrix
C. Trend Analysis

D. Scatter Plot

Q128. What is the primary goal of root cause analysis in incident investigations?

A. To identify individual blame

B. To uncover underlying systemic causes of incidents

C. To determine financial liability

D. To document rule violations

Q129. Which safety documentation provides evidence that all risk assessments, procedures, and audits are controlled and traceable?

A. Safety Record Management System

B. Hazard Communication Program

C. Inspection Schedule

D. Risk Register

Q130. What is the main benefit of conducting periodic management safety reviews?

A. To ensure regulatory reports are submitted

B. To align safety objectives with organizational strategy

C. To increase safety documentation volume

D. To enforce disciplinary measures

Q131. Which control strategy aims to eliminate hazards at the source before they reach workers?

A. Personal Protective Equipment

B. Engineering Control

C. Administrative Control

D. Elimination

Q132. What distinguishes proactive hazard reporting systems from traditional reactive systems?

A. They require management approval before reporting

B. They encourage early identification of risks before incidents occur

C. They record only major accidents

D. They operate only during audits

Q133. According to ICAO, what role does Safety Assurance play within the SMS framework?

A. Detecting and correcting deficiencies through monitoring and evaluation

B. Setting long-term safety policy

C. Conducting employee orientations

D. Managing financial risk

Q134. What is the primary objective of conducting safety performance trend analysis?

A. To monitor safety performance over time and predict future risks

B. To summarize financial costs

C. To compare competitors' performance

D. To review compliance with legal reporting

Q135. What does the term "safety maturity" refer to in organizational safety management?

A. The age of the safety program

B. The level of integration, consistency, and continuous improvement in safety culture

C. The number of audits performed annually

D. The size of the safety department

Q136. Which key factor differentiates a proactive safety program from a reactive one?

A. Reliance on incident data

B. Dependence on regulatory audits

C. Increased paperwork and reports

D. Focus on preventing risks before incidents occur

Q137. What is the purpose of safety objectives in a Safety Management System (SMS)?

A. To provide financial targets

B. To establish measurable goals that guide safety performance

C. To replace risk assessments
D. To determine employee bonuses
Q138. Which method best ensures employee involvement in safety program development?
A. Top-down communication only
B. Formation of joint safety committees
C. Delegating all responsibility to management
D. Conducting disciplinary reviews
Q139. Which principle ensures that safety responsibilities are clearly communicated throughout an organization?
A. Defined roles and accountability
B. Chain of command
C. Independent verification
D. Cross-training programs
Q140. What is the key outcome of implementing a robust hazard reporting system?
A. Increased disciplinary actions
B. Identification and mitigation of potential hazards before incidents occur
C. Simplified regulatory audits
D. Reduction in communication requirements
Q141. Which document establishes the foundational commitment of leadership to safety management principles?
A. Safety Objectives
B. Safety Policy
C. Risk Register
D. Incident Log
Q142. Which activity most directly supports continuous improvement in safety performance?
A. Periodic management reviews and corrective action tracking
B. Reducing training frequency
C. Outsourcing hazard assessments
D. Maintaining fixed procedures without revision
Q143. Which type of control aims to change how people work rather than alter the physical environment?
A. Administrative control
B. Engineering control
C. Elimination
D. Substitution
Q144. What is the most significant advantage of integrating employee feedback into safety program evaluation?
A. Reduces need for external audits
B. Provides practical insights into real-world conditions and control effectiveness
C. Increases administrative workload
D. Replaces performance metrics
Q145. What is the first step in the hierarchy of hazard controls?
A. Elimination
B. Engineering control
C. Substitution
D. Personal Protective Equipment
Q146. What key factor distinguishes effective safety leadership from simple management oversight?
A. Enforcement through penalties
B. Reducing documentation requirements
C. Motivating and influencing employees to prioritize safety values

D. Delegating accountability to subordinates

Q147. Which process ensures that accident trends and recurring issues are addressed strategically?

A. Hazard Identification

B. Safety Data Analysis

C. Root Cause Analysis

D. Safety Promotion

Q148. What is the ultimate goal of an incident reporting system in a safety management framework?

A. Assigning blame to responsible parties

B. Promoting learning and system improvement

C. Simplifying administrative reporting

D. Minimizing paperwork

Q149. According to the Safety Management System framework, what does the term "Safety Assurance" primarily verify?

A. That the organization meets production targets

B. That only management reports incidents

C. That employees attend training

D. That implemented safety processes are effective and compliant

Q150. What is the key advantage of implementing a near-miss reporting system?

A. It prevents legal liability

B. It identifies weaknesses before they result in incidents

C. It focuses on minor injuries only

D. It eliminates the need for hazard analysis

Q151. Which practice best promotes a culture of transparency within safety management?

A. Confidential reporting mechanisms

B. Publicly disclosing disciplinary actions

C. Restricting access to safety data

D. Centralizing communication only through management

Q152. Which aspect of safety communication ensures consistent messaging across the organization?

A. Informal conversations

B. Standardized communication protocols

C. Ad hoc announcements

D. Isolated department meetings

Q153. What is the main function of safety key performance indicators (KPIs)?

A. To replace audits

B. To measure, monitor, and improve safety performance over time

C. To eliminate human error

D. To satisfy financial reporting requirements

Q154. Which tool is most useful for evaluating the reliability of audit results among different auditors?

A. Inter-observer reliability assessment

B. Hazard analysis matrix

C. Root cause mapping

D. Job hazard checklist

Q155. Which factor most enhances safety accountability within an organization?

A. Reducing employee participation

B. Increasing supervision frequency

C. Assigning specific, measurable safety responsibilities to individuals

D. Outsourcing compliance duties

Q156. Which key factor ensures that corrective actions remain effective over time?

A. Regular verification and follow-up audits

B. Annual refresher training only
C. Increased paperwork
D. Restricting employee input
Q157. What is the primary purpose of benchmarking in a safety management context?
A. To compare safety performance with industry best practices
B. To replace risk assessments
C. To justify reduced training budgets
D. To standardize disciplinary measures
Q158. Which of the following is the most critical function of a Safety Management Review?
A. Reviewing employee attendance
B. Assessing system effectiveness and alignment with safety objectives
C. Scheduling social events
D. Recording incident reports
Q159. Which type of indicator provides early warning of potential safety issues?
A. Lagging indicator
B. Leading indicator
C. Financial indicator
D. Static performance metric
Q160. What is the key outcome of conducting a management of change (MOC) process?
A. Identification and control of new risks arising from changes
B. Reduction in administrative workload
C. Automatic compliance with all standards
D. Decrease in staff involvement
Q161. What is the ultimate goal of safety performance monitoring?
A. To predict and prevent incidents through data-driven insights
B. To increase report frequency
C. To satisfy auditing requirements
D. To simplify documentation
Q162. Which characteristic defines an effective safety culture?
A. Shared commitment to safety values at all organizational levels
B. Fear-driven compliance
C. Limited worker participation
D. Sole reliance on management decisions
Q163. Which document formally records lessons learned from incidents to prevent recurrence?
A. Emergency Plan
B. Safety Policy
C. Corrective Action Register
D. Equipment Manual
Q164. Which safety management process helps determine whether implemented controls are achieving desired outcomes?
A. Safety Assurance
B. Hazard Identification
C. Safety Policy
D. Risk Acceptance
Q165. What is the most important reason to conduct a Job Safety Analysis (JSA)?
A. To systematically identify hazards associated with specific tasks
B. To satisfy documentation audits
C. To assign worker penalties
D. To test employee performance
Q166. What is the purpose of establishing a hazard register?

A. To summarize audit budgets
B. To record employee schedules
C. To maintain training attendance
D. To document identified hazards and track mitigation actions
Q167. What ensures that risk control measures remain relevant as processes evolve?
A. Periodic risk reassessment and review
B. Fixed control implementation
C. Reduced documentation frequency
D. Single annual audit
Q168. What is the key purpose of establishing measurable safety targets?
A. To evaluate performance objectively and track progress
B. To simplify safety documentation
C. To reduce audit complexity
D. To increase employee workload
Q169. Which practice best demonstrates an organization's commitment to continuous safety improvement?
A. Maintaining unchanging safety manuals
B. Integrating findings from audits and feedback into revised procedures
C. Limiting employee participation in reviews
D. Conducting audits without follow-up
Q170. Which management action most effectively supports a positive safety culture?
A. Leading by example and visible involvement in safety activities
B. Delegating all safety duties
C. Issuing memos without participation
D. Avoiding employee communication
Q171. Which process helps determine whether a proposed project introduces unacceptable risk levels?
A. Risk Assessment
B. Safety Assurance
C. Emergency Preparedness
D. Safety Promotion
Q172. Which step follows hazard identification in the risk management process?
A. Risk assessment and prioritization
B. Training development
C. Policy approval
D. Performance reporting
Q173. Which element ensures that employees can perform their duties safely and competently?
A. Audit Review
B. Hazard Register
C. Training and Competency Development
D. Management of Change
Q174. What is the key purpose of safety performance reviews?
A. To evaluate achievements and identify opportunities for system enhancement
B. To document attendance
C. To enforce penalties
D. To summarize financial outcomes
Q175. Which factor is most crucial for ensuring effective communication of safety goals throughout the organization?
A. Clear, consistent, and multi-channel messaging
B. Technical jargon and lengthy reports

C. Exclusive management briefings
D. Restrictive communication practices

Q176. Which element is essential for ensuring safety data integrity within a management system?
A. Discretionary recordkeeping
B. Informal data collection
C. Verbal-only communication
D. Controlled documentation and version tracking

Q177. What is the main benefit of establishing a safety committee within an organization?
A. Promoting collaborative engagement between management and employees
B. Reducing leadership accountability
C. Eliminating the need for audits
D. Transferring compliance responsibility

Q178. Which factor most strongly contributes to the sustainability of a safety management system?
A. Continuous leadership involvement and accountability
B. Sporadic inspections
C. Isolated incident reviews
D. Temporary corrective actions

Q179. Which tool is most effective for identifying recurring issues in workplace incidents?
A. Training evaluation
B. Trend analysis
C. Job hazard checklist
D. Daily inspection logs

Q180. What is the key purpose of a Safety Management System manual?
A. To define policies, procedures, and responsibilities for implementing the SMS
B. To list employee salaries
C. To track training attendance
D. To summarize audit costs

Q181. Which activity ensures that identified safety hazards are effectively mitigated?
A. Implementation and verification of corrective actions
B. Conducting employee recognition programs
C. Scheduling performance appraisals
D. Filing incident reports without review

Q182. Which component of the PDCA cycle focuses on implementing planned safety measures?
A. Plan
B. Do
C. Check
D. Act

Q183. Which control strategy focuses on replacing a hazardous material or process with a safer alternative?
A. Elimination
B. Substitution
C. Engineering control
D. Administrative control

Q184. What is the main goal of implementing a Safety Performance Measurement program?
A. To track progress and verify safety management effectiveness
B. To reduce employee autonomy
C. To replace incident investigation
D. To increase administrative documentation

Q185. Which management function ensures that safety priorities are incorporated into organizational planning?

A. Document archiving
B. Post-incident reporting
C. Strategic integration of safety objectives
D. Resource reduction

Q186. What is the principal advantage of conducting behavioral safety observations?
A. Identifying and reinforcing safe work practices
B. Monitoring compliance only after incidents
C. Reducing employee communication
D. Limiting management participation

Q187. Which process evaluates whether implemented safety controls remain aligned with organizational objectives?
A. Management Review
B. Incident Investigation
C. Hazard Reporting
D. Risk Assessment

Q188. Which type of training is most effective for reinforcing safe behavior and preventing complacency?
A. Policy distribution without discussion
B. Initial onboarding only
C. One-time seminars
D. Regular refresher training

Q189. Which system element ensures transparency in tracking risk management activities?
A. Documentation and recordkeeping
B. Verbal briefings
C. Informal inspections
D. Confidential evaluations only

Q190. Which phase of the risk management process involves ranking hazards based on their likelihood and severity?
A. Control implementation
B. Risk evaluation and prioritization
C. Audit verification
D. Safety promotion

Q191. What is the purpose of a safety policy statement?
A. To communicate management's commitment and expectations regarding safety
B. To list all incidents
C. To track training hours
D. To assign disciplinary penalties

Q192. Which key performance indicator best measures proactive safety engagement?
A. Number of completed safety observations and near-miss reports
B. Number of lost-time injuries
C. Total medical claims
D. Insurance renewal rate

Q193. Which process ensures continuous improvement of safety controls based on operational feedback?
A. Hazard elimination only
B. Compliance certification
C. Performance review and corrective action
D. Documentation storage

Q194. What is the primary goal of an emergency preparedness program?
A. To ensure coordinated and effective response to incidents

B. To reduce document retention requirements
C. To assign disciplinary action post-event
D. To limit employee participation

Q195. What is the primary reason for maintaining consistent safety communication across all departments?

A. To ensure all employees receive uniform information and expectations
B. To reduce management workload
C. To simplify audit procedures
D. To minimize documentation

Q196. Which management activity ensures alignment between safety goals and organizational strategy?

A. Periodic management review meetings
B. Routine toolbox talks
C. Weekly inspection reports
D. Contractor evaluations

Q197. What is the most effective way to maintain employee engagement in safety programs?

A. Relying solely on audits
B. Increasing disciplinary actions
C. Limiting communication
D. Recognizing and rewarding proactive safety behavior

Q198. Which type of safety performance indicator reflects actual injury or incident outcomes?

A. Lagging indicator
B. Leading indicator
C. Predictive metric
D. Training index

Q199. What is the purpose of conducting internal safety audits?

A. To assess compliance and verify system effectiveness
B. To penalize departments for deficiencies
C. To replace external regulatory inspections
D. To reduce safety documentation

Q200. Which management principle is central to maintaining a strong safety culture?

A. Punitive enforcement
B. Delegation of accountability
C. Visible leadership commitment and consistency
D. Outsourced communication

Answer keys and Explanations (101-200)

Q101. Answer: B

Explanation: Training and Education ensure workers possess the knowledge, skills, and awareness to perform their duties safely. This element reinforces procedural understanding and builds competency through initial instruction, refreshers, and performance evaluations. Effective training fosters consistent behavior, hazard recognition, and adherence to best practices. By empowering workers with knowledge and confidence, organizations strengthen safety culture and reduce error-related risks across all operational levels.

Q102. Answer: B

Explanation: Leading and lagging indicators provide a balanced framework for evaluating safety performance. Leading indicators measure proactive actions like inspections and training, predicting potential improvements, while lagging indicators quantify past results, such as injuries or incidents. Together, they enable continuous monitoring, root-cause learning, and predictive control. This combination supports data-driven management, ensuring informed decisions that reinforce prevention and foster a cycle of continuous safety improvement.

Q103. Answer: B

Explanation: Safety Promotion focuses on developing a culture of safety awareness and continuous learning through effective communication, training, and leadership engagement. It ensures that everyone understands the organization's safety values and their role in achieving them. Through educational initiatives, sharing lessons learned, and reinforcing safety behaviors, Safety Promotion sustains motivation and commitment, embedding safety consciousness across all levels of the organization's workforce.

Q104. Answer: C

Explanation: The "Check" phase of the PDCA cycle involves evaluating and verifying whether implemented safety actions achieved intended outcomes. It relies on data collection, audits, and performance metrics to determine control effectiveness. This step identifies gaps and improvement opportunities, providing essential feedback for system refinement. By critically assessing results, organizations ensure that safety processes evolve responsively, maintaining alignment with goals and operational realities.

Q105. Answer: A

Explanation: Safety audits, as defined by VTT's research, systematically verify compliance with legal and procedural requirements and validate that safety systems are correctly implemented and functioning. This dual focus ensures that both the presence and effectiveness of safety measures are assessed. By combining compliance verification with operational validation, audits provide actionable insights that strengthen reliability, support continuous improvement, and uphold consistent safety performance across the organization.

Q106. Answer: A

Explanation: Management Review provides a structured forum for senior leaders to assess the effectiveness of the safety management system. It ensures that performance data, audit results, and emerging risks are analyzed to maintain alignment with organizational safety objectives. Through this process, leadership reaffirms commitment, allocates resources, and establishes strategic priorities. Regular management reviews sustain accountability and drive continual enhancement of safety governance and system maturity.

Q107. Answer: B

Explanation: A Hazard Register catalogs every identified workplace hazard, its assessed risk level, and associated control measures. It acts as a living document guiding risk management actions and facilitating compliance verification. By maintaining an updated register, organizations ensure transparency, traceability, and accountability in hazard control processes. It supports audits, reviews, and proactive hazard tracking, ensuring controls remain current and effective in mitigating operational risks.

Q108. Answer: A

Explanation: Top management holds ultimate accountability for defining measurable safety objectives that reflect organizational policy and strategic goals. Establishing such objectives provides direction, quantifiable targets, and performance indicators for all safety initiatives. Leadership uses these metrics to evaluate progress, allocate resources, and motivate continuous improvement. This high-level responsibility ensures safety remains a central business priority integrated into every strategic and operational decision.

Q109. Answer: C

Explanation: Engineering controls reduce risk by physically isolating workers from hazards through design or modification of equipment, processes, or facilities. Examples include ventilation systems, guarding, and noise enclosures. By addressing risks at their source, these controls minimize human dependency, providing reliable and consistent protection. Their preventive nature and permanence make engineering controls among the most effective interventions within the hierarchy of hazard management strategies.

Q110. Answer: B

Explanation: Safety inspections systematically identify hazardous conditions, unsafe behaviors, and noncompliance within the workplace. Their purpose is to detect and document issues before incidents occur, enabling prompt corrective actions. Regular inspections reinforce accountability, validate the effectiveness of controls, and maintain operational integrity. This proactive approach ensures continuous improvement, reduces risks, and sustains compliance with both internal standards and external regulatory requirements.

Q111. Answer: D

Explanation: Effective hazard communication relies on clear, comprehensive employee training that translates technical information into understandable, practical guidance. Workers must grasp labeling, pictograms, and Safety Data Sheet (SDS) content to recognize hazards and apply safe handling procedures. Consistent, accessible education empowers employees to act safely and respond effectively to emergencies, ensuring the communication program achieves its purpose—protecting health and promoting informed hazard awareness organization-wide.

Q112. Answer: B

Explanation: Effective safety leadership inspires others by modeling the desired safety behaviors and attitudes. Leaders demonstrate commitment through consistent actions, communication, and involvement in safety initiatives. This approach builds trust, empowers participation, and reinforces shared responsibility. Inspirational leadership transcends rule enforcement, cultivating intrinsic motivation and a collective mindset that values safety as an organizational priority and integral component of every operational decision.

Q113. Answer: A

Explanation: Continuous improvement sustains safety management systems by transforming feedback, performance data, and lessons learned into actionable changes. This iterative process refines policies, enhances training, and strengthens controls to adapt to emerging risks. By embedding review and adaptation cycles, organizations maintain relevance and resilience. Continuous improvement transforms compliance-based safety into a dynamic, evolving system that consistently enhances worker protection and operational reliability.

Q114. Answer: B

Explanation: Management of Change (MOC) ensures that modifications to processes, equipment, or personnel undergo thorough safety evaluation prior to implementation. This process identifies new hazards, assesses potential impacts, and defines control measures to prevent unintended consequences. MOC preserves safety integrity by requiring review, authorization, and documentation, ensuring that operational evolution occurs within a controlled, risk-aware framework aligned with the organization's overall safety objectives.

Q115. Answer: B

Explanation: Performance feedback reinforces learning by closing the communication loop between management and employees. It allows individuals and teams to understand performance results, recognize achievements, and identify areas for growth. Constructive feedback fosters engagement, accountability, and transparency, translating performance data into behavioral change. This ongoing dialogue ensures lessons learned from incidents, audits, or training lead to meaningful, continuous improvement throughout the safety management system.

Q116. Answer: D

Explanation: The successful implementation of a safety management system relies on visible management commitment coupled with active employee engagement. When leadership demonstrates genuine dedication through resource allocation, communication, and example-setting, it reinforces a culture of accountability. Employee involvement ensures that safety processes reflect operational realities. Together, these elements cultivate ownership, shared responsibility, and continuous improvement, transforming safety principles into daily operational habits that sustain organizational excellence.

Q117. Answer: B

Explanation: Safety performance indicators measure how effectively the safety management system achieves its objectives. They provide quantifiable evidence of progress, revealing strengths, weaknesses, and trends across operations. By tracking both proactive and reactive measures, organizations identify emerging risks and improvement opportunities. This continuous measurement process supports data-driven decision-making, strengthens accountability, and ensures that safety performance aligns with organizational goals and regulatory expectations.

Q118. Answer: B

Explanation: A just culture promotes fairness by distinguishing between acceptable human errors and negligent behaviors. It encourages open reporting by ensuring that employees will not be punished for unintentional mistakes, fostering trust and transparency. This approach transforms incidents into learning opportunities, enabling organizations to identify root causes and strengthen systems. Balancing accountability with compassion cultivates responsibility while maintaining psychological safety necessary for continuous safety improvement.

Q119. Answer: C

Explanation: Risk assessment systematically identifies potential hazards, evaluates their likelihood and consequences, and prioritizes them for mitigation. This structured process provides the foundation for informed decision-making in hazard control. By quantifying risk, organizations can allocate resources efficiently and select controls proportionate to the level of threat. Risk assessment transforms safety from reactive problem-solving into a proactive management discipline focused on prevention and continuous protection.

Q120. Answer: B

Explanation: Integrating safety with quality and environmental management streamlines overlapping functions such as auditing, training, and continuous improvement. Shared processes reduce redundancy and enhance communication across departments. This holistic approach fosters organizational efficiency, aligning safety goals with broader performance objectives. By managing risks in tandem with quality and environmental standards, organizations build synergy that strengthens compliance, sustainability, and long-term operational excellence.

Q121. Answer: A

Explanation: Safety Assurance ensures continuous learning and system improvement by monitoring performance and verifying corrective actions. It evaluates incident outcomes, audits, and safety data to confirm that remedial measures are both applied and effective. This element sustains feedback loops essential for organizational learning. Through systematic evaluation and documentation, Safety Assurance enhances accountability, strengthens resilience, and drives a proactive culture of continuous improvement.

Q122. Answer: B

Explanation: Residual risk represents the level of danger that remains after all reasonable mitigation strategies have been applied. It acknowledges that absolute elimination of risk is impractical, emphasizing continuous monitoring and periodic reassessment. Understanding residual risk allows management to make informed decisions about risk acceptance, additional controls, or operational changes. It is a core concept in maintaining realistic, data-informed safety strategies aligned with organizational tolerance levels.

Q123. Answer: D

Explanation: Consultation and Participation engage employees in developing, implementing, and improving occupational health and safety practices. This process empowers workers to contribute firsthand knowledge of operational risks and propose effective solutions. It builds ownership and ensures that risk control measures are practical and widely accepted. Active participation fosters collaboration between workers and management, leading to more robust, inclusive, and sustainable safety systems.

Q124. Answer: B

Explanation: Standard Operating Procedures (SOPs) define uniform methods for executing tasks safely and efficiently. They ensure consistency, reduce variability, and prevent unsafe practices by establishing clear expectations. SOPs serve as training references and compliance tools, helping employees understand and follow best practices. By standardizing operations, organizations minimize risk exposure and enhance performance reliability across diverse work environments and personnel levels.

Q125. Answer: B

Explanation: Safety Promotion encompasses communication, training, and awareness programs that instill safety values and reinforce organizational commitment. It ensures that safety messages, lessons learned, and best practices are shared across all levels. By nurturing a positive safety culture, Safety Promotion builds trust, encourages open reporting, and enhances motivation. This continuous engagement transforms safety into a shared organizational value rather than a procedural requirement.

Q126. Answer: C

Explanation: Management of Change (MOC) systematically evaluates proposed changes to processes, materials, or personnel for potential safety impacts before implementation. This proactive approach prevents introducing unforeseen hazards by requiring review, documentation, and approval. MOC ensures that risk assessments, control measures, and communication plans are completed, maintaining operational stability. It supports informed decision-making and safeguards against disruptions to established safety controls.

Q127. Answer: B

Explanation: A Risk Matrix provides a visual framework for assessing hazards by combining their likelihood and potential severity. It helps prioritize risks, guiding management toward allocating resources efficiently. This structured tool converts qualitative judgments into clear, actionable categories, simplifying communication among stakeholders. By ranking risks systematically, the matrix supports consistent evaluation and informed control decisions within the organization's broader safety management strategy.

Q128. Answer: B

Explanation: Root cause analysis identifies the fundamental system or process deficiencies that allow incidents to occur. It shifts focus from individual errors to organizational factors such as training gaps, inadequate controls, or procedural weaknesses. By addressing these root causes, organizations prevent recurrence and strengthen systemic resilience. The outcome promotes sustainable improvement by embedding lessons learned into management systems and operational practices.

Q129. Answer: A

Explanation: A Safety Record Management System organizes and preserves documentation related to risk assessments, audits, and procedural controls. It ensures data integrity, traceability, and accessibility for verification and continuous improvement. Proper record management supports compliance, demonstrates due diligence, and enables informed decision-making. This systematic approach enhances transparency, accountability, and reliability across all elements of the safety management system.

Q130. Answer: B

Explanation: Periodic management safety reviews ensure that safety objectives remain aligned with organizational goals and external conditions. They allow senior leadership to evaluate performance data, assess emerging risks, and determine whether resources are sufficient. This top-down review reinforces accountability and adaptability, ensuring that the safety management system evolves with the organization's strategic direction and continues delivering measurable, lasting improvements.

Q131. Answer: D

Explanation: Elimination removes the hazard completely from the workplace, preventing any potential exposure. As the most effective risk control measure, it addresses danger at its root rather than mitigating effects. By redesigning processes or substituting tasks, elimination provides a permanent solution that safeguards workers without reliance on behavioral controls. This approach epitomizes proactive, preventive risk management integral to advanced safety systems.

Q132. Answer: B

Explanation: Proactive hazard reporting systems empower employees to report unsafe conditions and near-misses early, allowing corrective actions before harm occurs. This approach builds a culture of openness and prevention, contrasting with reactive systems that respond only after incidents. Encouraging timely reports enhances learning, strengthens risk control, and fosters collective responsibility for continuous safety improvement throughout the organization.

Q133. Answer: A

Explanation: Safety Assurance functions as the feedback mechanism within the Safety Management System, ensuring that implemented controls perform as expected. Through audits, monitoring, and evaluation, it identifies deficiencies, verifies compliance, and confirms that corrective actions achieve desired outcomes. This process promotes accountability and continuous improvement, ensuring operational safety integrity remains consistent with organizational standards and regulatory expectations.

Q134. Answer: A

Explanation: Safety performance trend analysis tracks safety data over time to identify patterns, deviations, and emerging risks. By understanding trends, organizations can predict potential issues before they escalate and adjust preventive measures accordingly. This analytical process enhances decision-making, supports continuous improvement, and reinforces proactive management by transforming historical information into actionable insight for strategic risk prevention.

Q135. Answer: B

Explanation: Safety maturity reflects how deeply safety principles are embedded within an organization's culture and operations. High maturity indicates consistent leadership commitment, employee participation, and adaptive learning processes. It demonstrates a transition from compliance-driven practices to proactive risk management. Measuring safety maturity helps organizations assess cultural development and prioritize initiatives that enhance resilience, accountability, and sustainable safety performance.

Q136. Answer: D

Explanation: A proactive safety program focuses on identifying and controlling hazards before incidents occur. It emphasizes preventive strategies such as hazard reporting, risk assessment, and predictive analytics. This forward-looking approach cultivates a culture of anticipation and readiness rather than reaction. By prioritizing prevention, organizations reduce injury rates, improve efficiency, and demonstrate mature safety management practices that sustain operational reliability and employee confidence.

Q137. Answer: B

Explanation: Safety objectives provide measurable, time-bound goals aligned with the organization's safety policy and risk management priorities. They serve as benchmarks for performance evaluation, enabling management to assess progress and allocate resources effectively. These objectives transform strategic safety commitments into actionable plans, fostering accountability and continuous improvement. Well-defined safety objectives help maintain focus, motivate personnel, and demonstrate leadership's commitment to achieving sustainable safety excellence.

Q138. Answer: B

Explanation: Joint safety committees ensure structured participation by employees in identifying hazards, proposing solutions, and evaluating safety initiatives. This collaboration bridges communication between management and workers, allowing practical insights to shape policy. Employee representation builds ownership, fosters transparency, and encourages continuous engagement. Committees serve as active feedback channels, strengthening both morale and safety culture through collective responsibility and mutual trust within the organization.

Q139. Answer: A

Explanation: Clearly defining roles and accountability ensures that everyone understands their specific safety responsibilities and reporting lines. This principle prevents ambiguity, reduces overlap, and enhances coordination. When expectations are documented and communicated, individuals take ownership of tasks, while management maintains oversight. Such clarity ensures consistent implementation of controls, efficient resource use, and improved safety performance by aligning actions across all organizational levels.

Q140. Answer: B

Explanation: A strong hazard reporting system encourages employees to report unsafe conditions and near misses, enabling early detection and correction of potential hazards. This proactive process transforms employee observations into actionable insights that prevent incidents. It also promotes open communication and trust, essential elements of a positive safety culture. Ultimately, such systems foster continuous improvement and significantly reduce workplace risk exposure.

Q141. Answer: B

Explanation: The Safety Policy articulates leadership's dedication to integrating safety into every aspect of operations. It serves as the cornerstone of the Safety Management System by defining principles, expectations, and accountability for all personnel. This declaration aligns the organization's safety philosophy with its strategic mission, ensuring that management commitment is visible, measurable, and continuously reinforced throughout the organization's daily practices.

Q142. Answer: A

Explanation: Continuous improvement is achieved through regular management reviews that evaluate safety performance and verify implementation of corrective actions. This process ensures that lessons learned from incidents and audits translate into meaningful change. Tracking corrective measures maintains accountability and encourages a learning culture. These reviews validate that objectives are met and that the safety system evolves to address emerging risks effectively.

Q143. Answer: A

Explanation: Administrative controls manage risk by adjusting work practices, schedules, or procedures instead of physically modifying the environment. Examples include job rotation, safe work instructions, and training. These controls reduce exposure time or likelihood by guiding behavior through policy and supervision. While less effective than engineering solutions, administrative controls remain essential in ensuring consistent compliance and minimizing residual risks within the workplace.

Q144. Answer: B

Explanation: Employee feedback delivers firsthand, practical perspectives on how safety controls function in actual work conditions. Workers can identify procedural gaps, hidden hazards, or inefficiencies that formal assessments may overlook. Integrating this feedback ensures solutions are both realistic and effective. It also fosters ownership and engagement, reinforcing a safety culture based on mutual respect and shared accountability between management and workforce.

Q145. Answer: A

Explanation: Elimination is the first and most effective step in the hierarchy of controls, aiming to completely remove the hazard from the workplace. By redesigning tasks, processes, or materials to remove danger sources, elimination ensures permanent risk reduction. This proactive approach addresses hazards at their origin, preventing exposure altogether and serving as the foundation for all subsequent control strategies.

Q146. Answer: C

Explanation: Effective safety leadership motivates through influence, not authority. Leaders demonstrate visible commitment, communicate consistently, and embody safety values in their actions. This approach inspires employees to take personal responsibility for safety, fostering trust and participation. Leadership focuses on guiding, supporting, and empowering teams, transforming compliance into intrinsic motivation that drives sustainable, organization-wide safety excellence.

Q147. Answer: B

Explanation: Safety Data Analysis examines collected incident and near-miss data to identify recurring patterns or emerging trends. This process enables organizations to target systemic weaknesses rather than isolated events. By transforming raw data into actionable insights, safety data analysis supports resource prioritization and strategic risk management decisions, strengthening the overall efficiency and responsiveness of the safety program.

Q148. Answer: B

Explanation: The goal of an incident reporting system is to promote organizational learning and drive continuous improvement. It transforms reported data into preventive actions, allowing the identification of patterns, hazards, and root causes. A non-punitive environment encourages honest reporting, ensuring lessons learned enhance future controls. This fosters transparency, accountability, and a proactive culture focused on safety enhancement rather than punishment.

Q149. Answer: D

Explanation: Safety Assurance confirms that implemented processes achieve desired safety outcomes and comply with regulations. It involves performance monitoring, audits, and data analysis to ensure systems function as designed. By validating control effectiveness and ensuring accountability, Safety Assurance maintains confidence in safety programs and guides corrective action planning for continual improvement.

Q150. Answer: B

Explanation: A near-miss reporting system captures early warning signals of potential incidents, providing opportunities for proactive intervention. These reports allow organizations to identify vulnerabilities in processes or behaviors before they cause harm. By analyzing near-miss data, management can strengthen preventive measures, fostering a learning culture that values transparency and continuous risk reduction across all operational levels.

Q151. Answer: A

Explanation: Confidential reporting mechanisms encourage employees to share safety concerns without fear of reprisal. This trust-based system improves hazard identification and supports early intervention. By ensuring privacy, organizations create a psychologically safe environment that fosters openness and collaboration. Transparency through confidential reporting strengthens collective accountability and enhances the reliability of safety data collection and analysis.

Q152. Answer: B

Explanation: Standardized communication protocols establish clear procedures for distributing safety information across departments. Consistent language, frequency, and format ensure all employees receive accurate and timely updates. This standardization reduces misunderstandings, supports compliance, and enhances alignment with organizational safety objectives. Structured communication builds unity, clarity, and efficiency in conveying essential safety messages throughout all levels of the organization.

Q153. Answer: B

Explanation: Safety Key Performance Indicators (KPIs) measure and monitor progress toward defined safety objectives. They transform data into performance insights, helping management identify trends and evaluate the impact of safety initiatives. Effective KPIs enable benchmarking, support evidence-based decision-making, and drive continuous improvement. Regular analysis of these metrics ensures that safety performance remains transparent, measurable, and aligned with strategic organizational goals.

Q154. Answer: A

Explanation: An inter-observer reliability assessment evaluates the consistency of findings among multiple auditors assessing the same system. High reliability ensures that conclusions are objective and not influenced by individual interpretation. This assessment validates the credibility and reproducibility of audit processes, confirming that the methodology and criteria used produce uniform, trustworthy outcomes across diverse evaluators.

Q155. Answer: C

Explanation: Assigning specific, measurable safety responsibilities ensures individuals understand their role in achieving safety objectives. Clear accountability fosters ownership and commitment, linking personal performance to organizational outcomes. When employees are aware of expectations and consequences, engagement rises, and compliance improves. Defined responsibilities create a structured framework that supports consistent behavior, efficient oversight, and stronger alignment with the organization's safety management system.

Q156. Answer: A

Explanation: Regular verification and follow-up audits confirm that corrective actions implemented after incidents or inspections achieve their intended purpose. Continuous review ensures that the controls remain functional and relevant despite operational changes. This practice supports accountability and sustainability by identifying potential lapses early. Consistent verification prevents regression and reinforces the culture of ongoing improvement essential to a mature safety management system.

Q157. Answer: A

Explanation: Benchmarking allows organizations to compare their safety performance and management processes with industry leaders or established standards. This analysis reveals gaps and improvement opportunities while highlighting effective practices that can be adopted. Benchmarking supports informed decision-making, fosters innovation, and helps build competitive advantage through superior safety culture and operational excellence aligned with proven, data-driven methodologies.

Q158. Answer: B

Explanation: A Safety Management Review evaluates the performance and effectiveness of the organization's safety management system. It ensures that objectives are met, identifies emerging risks, and aligns safety initiatives with strategic goals. This structured review, led by senior management, reinforces leadership commitment, promotes transparency, and ensures continuous adaptation of the system to operational realities and future safety challenges.

Q159. Answer: B

Explanation: Leading indicators measure proactive activities such as inspections, training completion, and hazard reporting. They identify early signs of system weakness before incidents occur, allowing preventive action. Unlike lagging indicators, which reflect outcomes, leading indicators guide continuous improvement by emphasizing behaviors and processes that strengthen risk management and build long-term safety resilience within the organization.

Q160. Answer: A

Explanation: The Management of Change (MOC) process systematically identifies and evaluates risks associated with operational, procedural, or organizational modifications. By analyzing potential safety implications before implementation, it prevents new hazards from emerging. This proactive approach protects system integrity, ensures continuity, and promotes accountability through structured documentation and review before approving any change.

Q161. Answer: A

Explanation: Safety performance monitoring gathers and evaluates safety data to identify patterns, measure success, and forecast potential issues. This proactive process converts data into actionable insights, enabling preventive decisions before incidents occur. Continuous monitoring enhances transparency, accountability, and strategic planning by ensuring safety objectives remain relevant and effective across all levels of the organization.

Q162. Answer: A

Explanation: An effective safety culture exists when safety is recognized as a shared value embraced by leadership and workers alike. It emphasizes open communication, mutual respect, and proactive engagement in hazard management. Employees feel empowered to identify risks, report concerns, and participate in solutions. This unity of purpose fosters trust, accountability, and consistent adherence to safety practices across the organization.

Q163. Answer: C

Explanation: The Corrective Action Register captures identified deficiencies, corresponding solutions, and lessons learned from incidents or audits. It ensures that each corrective measure is tracked from identification through verification of completion. This document preserves institutional knowledge and promotes accountability by ensuring systemic issues are addressed and prevented from recurring, reinforcing a cycle of continuous organizational improvement.

Q164. Answer: A

Explanation: Safety Assurance verifies the ongoing effectiveness of safety controls through performance monitoring, audits, and reviews. It ensures that risk mitigation strategies deliver expected results and identifies areas for refinement. This continuous validation process strengthens confidence in safety systems and supports adaptive management capable of responding effectively to evolving operational conditions.

Q165. Answer: A

Explanation: A Job Safety Analysis (JSA) breaks tasks into steps to identify hazards and define controls for each. This process ensures that preventive measures are built into work procedures, reducing the risk of human error or injury. JSAs empower workers with understanding and ownership of safe methods, making it a cornerstone of proactive, task-specific safety management.

Q166. Answer: D

Explanation: A hazard register compiles all recognized workplace hazards, their assessed risk levels, and corresponding control measures. It provides a living record that guides decision-making and facilitates regular review. Maintaining this documentation promotes accountability, ensures visibility of residual risks, and supports compliance with safety standards by tracking mitigation progress over time.

Q167. Answer: A

Explanation: Periodic risk reassessment ensures that implemented controls continue to match operational realities. As technology, materials, or workflows change, reassessment detects emerging risks or inefficiencies. Regular review ensures continual adaptation, maintaining optimal control effectiveness. This proactive approach supports ongoing compliance and resilience within an evolving safety management environment.

Q168. Answer: A

Explanation: Measurable safety targets provide a quantifiable basis for tracking performance over time. They allow management to gauge progress toward objectives, identify trends, and prioritize interventions. Clear metrics transform abstract goals into actionable benchmarks, promoting accountability and transparency. Consistent evaluation against these targets ensures continuous improvement and sustained focus on achieving safety excellence.

Q169. Answer: B

Explanation: Continuous improvement requires integrating lessons learned from audits, incident reviews, and employee feedback into procedural updates. This ensures that safety systems evolve based on real performance data rather than static assumptions. Active integration of findings enhances system adaptability, supports innovation, and demonstrates a sustained commitment to protecting workers and improving operational reliability.

Q170. Answer: A

Explanation: Leadership that demonstrates visible involvement—participating in safety meetings, inspections, and discussions—signals genuine commitment. This behavior reinforces accountability and motivates workers to adopt safety as a shared value. Leading by example fosters trust, engagement, and ownership, creating a culture where safety is prioritized at every level of the organization.

Q171. Answer: A

Explanation: Risk Assessment identifies potential hazards within proposed projects, evaluating their probability and consequences to determine acceptability. This analysis ensures that new initiatives proceed safely and within the organization's risk tolerance. Performing assessments early allows for design modifications and preventive controls, ensuring informed, responsible decision-making before implementation.

Q172. Answer: A

Explanation: After identifying hazards, risk assessment and prioritization determine which risks require immediate control and which can be monitored. This process evaluates likelihood and severity, guiding resource allocation toward high-risk areas. Structured prioritization ensures efficient and focused management, enabling targeted interventions that deliver the highest safety impact.

Q173. Answer: C

Explanation: Training and Competency Development equip employees with the knowledge and skills required to perform tasks safely. Ongoing instruction, verification, and refresher programs maintain competency levels. This investment enhances safety awareness, improves job performance, and minimizes the risk of error. Competency-based training demonstrates commitment to continuous learning and safe operational excellence.

Q174. Answer: A

Explanation: Safety performance reviews assess whether objectives are being met and determine opportunities for improvement. By analyzing data trends, incident reports, and feedback, management ensures that safety programs remain effective. These reviews promote transparency, accountability, and adaptability, fostering a continuous learning environment where safety outcomes consistently improve.

Q175. Answer: A

Explanation: Effective communication of safety goals depends on clarity, consistency, and accessibility across all communication channels. Using understandable language and regular reinforcement ensures alignment between leadership intent and workforce understanding. Multi-channel messaging—through meetings, signage, and digital tools—strengthens engagement, keeps safety priorities visible, and promotes unified action toward shared objectives.

Q176. Answer: D

Explanation: Controlled documentation and version tracking maintain data accuracy and prevent unauthorized modification or loss. This process ensures that only the most recent, verified safety information guides decisions. Version control also supports traceability and regulatory compliance, allowing organizations to demonstrate accountability. Maintaining structured records strengthens data integrity, supports audits, and provides a reliable foundation for continuous safety improvement.

Q177. Answer: A

Explanation: A safety committee fosters collaboration by bringing management and employees together to identify hazards, review incidents, and propose improvements. This partnership ensures that both perspectives influence decisions, promoting practical and effective solutions. Regular meetings reinforce transparency, mutual trust, and shared ownership of safety outcomes. By engaging all levels, committees build cultural strength and continuous improvement momentum across the organization.

Q178. Answer: A

Explanation: Sustainability in safety management depends on consistent leadership involvement. When management remains visibly engaged and accountable, safety values are reinforced throughout the organization. Leadership ensures resources, recognition, and priorities align with safety objectives. Continuous oversight embeds safety into operational decisions, preventing stagnation and ensuring long-term performance improvement through active governance and example-setting.

Q179. Answer: B

Explanation: Trend analysis identifies patterns across multiple incidents or near misses, highlighting recurring causes and systemic weaknesses. This data-driven tool transforms historical records into predictive insights that guide preventive strategies. By analyzing trends, organizations can prioritize resources, address underlying issues, and prevent future incidents. It is an essential mechanism for continuous safety improvement and risk mitigation.

Q180. Answer: A

Explanation: The Safety Management System manual serves as a comprehensive reference document outlining the organization's safety policies, processes, and responsibilities. It ensures consistency, clarity, and compliance by defining how each element of the SMS is implemented and maintained. The manual supports employee understanding, provides operational guidance, and acts as the central framework for continuous safety performance management.

Q181. Answer: A

Explanation: Effective hazard mitigation requires implementation of corrective actions followed by verification to confirm their success. This dual-step process transforms recommendations into tangible safety improvements and prevents recurrence of similar issues. Verification ensures the chosen controls remain functional and effective over time, supporting accountability and maintaining confidence in the safety system's reliability.

Q182. Answer: B

Explanation: The "Do" phase of the Plan-Do-Check-Act cycle involves executing planned actions and safety improvements. During this stage, organizations implement control measures, conduct training, and establish procedures to address identified risks. Successful execution ensures that theoretical planning becomes operational reality, setting the foundation for measurement and feedback in subsequent phases of the safety improvement cycle.

Q183. Answer: B

Explanation: Substitution reduces risk by replacing a hazardous material, process, or equipment with a less hazardous alternative. This control strategy minimizes exposure potential without compromising functionality. It exemplifies proactive risk reduction within the hierarchy of controls, addressing hazards at their source. Effective substitution requires thorough assessment to ensure new risks are not introduced during implementation.

Q184. Answer: A

Explanation: Safety Performance Measurement evaluates how effectively an organization's safety initiatives achieve their intended outcomes. By tracking performance indicators, it provides objective data for decision-making and identifies areas requiring improvement. This structured approach enhances accountability, transparency, and continuous refinement of safety systems, ensuring consistent progress toward defined objectives and improved operational safety performance.

Q185. Answer: C

Explanation: Strategic integration of safety objectives embeds safety considerations into long-term business planning. This ensures that decisions about production, budgets, and workforce development account for safety implications. Integrating safety at the strategic level demonstrates leadership commitment, aligns resources with priorities, and creates an environment where safety becomes inseparable from operational excellence.

Q186. Answer: A

Explanation: Behavioral safety observations proactively monitor and encourage safe practices in real-time. Observers provide feedback that reinforces desired behaviors and corrects unsafe actions. This positive reinforcement strengthens personal accountability and promotes a culture of mutual support. Regular observations help prevent incidents, bridge communication gaps, and foster consistent adherence to safety standards.

Q187. Answer: A

Explanation: Management Review examines performance metrics, audit results, and compliance data to confirm alignment between safety controls and strategic goals. This oversight ensures the system evolves to address emerging risks and changing priorities. By maintaining alignment, management reviews preserve system relevance, support accountability, and promote sustained improvement in safety outcomes.

Q188. Answer: D

Explanation: Regular refresher training reinforces existing knowledge, updates employees on new hazards, and combats complacency. By continuously engaging workers, it ensures that safe behaviors remain habitual and aligned with evolving procedures. Refresher sessions encourage dialogue, strengthen confidence, and promote consistent adherence to safety standards throughout the workforce.

Q189. Answer: A

Explanation: Documentation and recordkeeping provide traceability of all risk management actions, from hazard identification to control verification. This transparency ensures accountability and facilitates audits or reviews. Proper documentation supports decision-making, demonstrates compliance, and enables lessons learned to be shared organization-wide for sustained improvement and knowledge retention.

Q190. Answer: B

Explanation: Risk evaluation and prioritization assess identified hazards by combining likelihood and severity to determine significance. This ranking guides where resources should be focused for maximum impact. Prioritization ensures that high-risk issues receive immediate attention while maintaining proportional control efforts, optimizing safety investments, and enhancing strategic risk management.

Q191. Answer: A

Explanation: A safety policy statement articulates leadership's dedication to providing a safe workplace and outlines fundamental safety objectives. It sets the tone for the organization's culture and establishes accountability at every level. This visible declaration reinforces the organization's intent to prioritize safety, guiding all related activities and decision-making processes.

Q192. Answer: A

Explanation: Tracking completed safety observations and near-miss reports demonstrates proactive engagement by employees and supervisors. This indicator reflects organizational participation in hazard prevention rather than reaction to incidents. High reporting activity signals a strong safety culture, active communication, and commitment to continuous improvement through early risk identification and mitigation.

Q193. Answer: C

Explanation: Performance review and corrective action convert feedback, performance data, and audit findings into tangible system improvements. This cyclical process ensures that controls evolve alongside operations, sustaining effectiveness. Continuous review fosters accountability and encourages innovation, embedding improvement into the organization's safety management framework.

Q194. Answer: A

Explanation: Emergency preparedness ensures that personnel are trained, equipped, and coordinated to respond effectively to emergencies. It minimizes harm, property loss, and downtime through clear procedures, communication protocols, and periodic drills. A robust program builds confidence and resilience, ensuring rapid and organized response that safeguards people and operations.

Q195. Answer: A

Explanation: Consistent safety communication guarantees that all employees understand safety expectations, updates, and performance goals. Uniform messaging prevents confusion, supports compliance, and reinforces shared responsibility. When communication is coordinated across departments, safety priorities remain aligned, strengthening collaboration and sustaining a unified, informed workforce.

Q196. Answer: A

Explanation: Periodic management review meetings allow leadership to assess safety performance against organizational objectives and realign strategies as needed. This activity reinforces accountability, identifies resource requirements, and ensures that safety priorities remain integral to the business mission. By maintaining regular oversight, management strengthens commitment and fosters continuous improvement throughout the safety management system.

Q197. Answer: D

Explanation: Recognition of proactive safety behavior motivates employees to take ownership of safety initiatives. Positive reinforcement builds morale, encourages participation, and demonstrates that management values safe conduct. Reward systems transform compliance into intrinsic motivation, leading to consistent engagement and stronger adherence to safety practices across all organizational levels.

Q198. Answer: A

Explanation: Lagging indicators measure reactive outcomes such as injury rates, lost-time incidents, and property damage. They provide retrospective insight into system performance and help validate whether existing controls effectively prevent harm. Tracking these outcomes supports benchmarking and drives targeted corrective actions to improve future safety performance and risk mitigation.

Q199. Answer: A

Explanation: Internal safety audits evaluate adherence to internal policies and external standards, ensuring that safety programs operate effectively. They identify gaps, assess control performance, and promote transparency. By providing objective feedback, audits guide continual improvement efforts and strengthen the organization's ability to maintain compliance and operational integrity.

Q200. Answer: C

Explanation: Visible leadership commitment and consistency demonstrate that safety is a core organizational value. When leaders actively participate in safety activities, communicate openly, and allocate resources appropriately, they set a standard for others to follow. This consistency builds trust and embeds safety as an integral part of daily operations and decision-making.

Questions (201-300)

Q201. Which document defines the organization's framework for implementing, maintaining, and improving its safety program?

A. Safety Manual
B. Audit Checklist
C. Risk Register
D. Incident Log

Q202. What is the primary objective of a hazard identification program?

A. To review past performance data only
B. To systematically recognize potential sources of harm before incidents occur
C. To evaluate employee absenteeism
D. To document existing hazards without action

Q203. What is the main purpose of safety key performance indicators (KPIs)?

A. To measure and improve progress toward safety objectives
B. To track employee absenteeism
C. To calculate annual bonuses
D. To replace incident reporting

Q204. Which management process ensures safety remains a core component of daily decision-making?

A. Integration of safety into business planning
B. Periodic financial audits
C. Contractor prequalification
D. Document retention

Q205. What is the function of incident trend analysis in a safety management system?

A. To increase documentation load
B. To record single-event causes
C. To identify recurring patterns and implement preventive measures
D. To satisfy legal formality

Q206. Which term describes the level of risk remaining after control measures have been applied?

A. Residual risk
B. Baseline risk
C. Primary risk
D. Hidden hazard

Q207. Which process identifies safety gaps through systematic comparison against set standards?

A. Gap analysis
B. Risk mapping
C. Trend observation
D. Hazard ranking

Q208. What ensures consistency and accountability in corrective action management?

A. Independent contractor assessment
B. Informal communication
C. Annual review only
D. Action tracking and verification system

Q209. Which method most effectively promotes learning from safety incidents?

A. Conducting root cause analyses and sharing lessons learned
B. Limiting access to reports
C. Implementing punitive measures
D. Archiving incidents without review

Q210. What is the purpose of conducting safety culture assessments?

A. To evaluate organizational attitudes, values, and behaviors toward safety
B. To measure financial performance
C. To test compliance with environmental laws
D. To determine audit costs
Q211. Which system ensures that safety processes continue to function effectively between audits?
A. Contractor review
B. Annual safety survey
C. Continuous monitoring and internal verification
D. Post-incident analysis
Q212. Which factor best promotes a culture of trust in safety reporting?
A. A non-punitive reporting policy
B. Strict disciplinary enforcement
C. Confidentiality limitations
D. Infrequent communication
Q213. What is the purpose of a management system maturity model in safety programs?
A. To evaluate the development level and effectiveness of the system
B. To replace audits
C. To limit communication
D. To track individual performance
Q214. What is the core objective of performance feedback within a safety management system?
A. To assign blame
B. To reinforce learning and motivate continuous improvement
C. To document disciplinary actions
D. To track absenteeism
Q215. Which factor most enhances the credibility of safety data analysis?
A. Accuracy, validation, and transparency of data sources
B. Limited access to information
C. Manual recordkeeping only
D. Use of anecdotal reports
Q216. Which process ensures that safety-related decisions are based on accurate and timely data?
A. Equipment calibration
B. Hazard observation
C. Safety data management and analysis
D. Employee interviews
Q217. What is the most effective method to verify that corrective actions have been successfully implemented?
A. Conducting follow-up audits and performance evaluations
B. Holding staff meetings
C. Updating documentation only
D. Issuing reminders
Q218. Which element of a safety management system focuses on fostering organizational learning?
A. Safety Promotion
B. Safety Assurance
C. Safety Policy
D. Risk Assessment
Q219. What is the main purpose of safety performance evaluation?
A. To summarize audit costs
B. To record disciplinary actions
C. To document training attendance
D. To assess the effectiveness of safety measures and identify improvement areas

Q220. What is the primary function of an employee participation program in safety management?
A. To involve workers in identifying hazards and developing solutions
B. To reduce management accountability
C. To replace audits
D. To manage finances
Q221. Which aspect of safety leadership has the greatest influence on employee behavior?
A. Demonstrating consistent commitment to safety through action
B. Enforcing strict penalties
C. Delegating safety tasks
D. Issuing written policies only
Q222. What is the role of management review in a safety management system?
A. To replace incident investigation
B. To evaluate system performance and ensure alignment with strategic goals
C. To manage administrative costs
D. To verify employee attendance
Q223. Which safety control method focuses on altering the work environment to reduce exposure to hazards?
A. Engineering control
B. Administrative control
C. PPE
D. Elimination
Q224. What is the main advantage of proactive hazard reporting?
A. Early detection and mitigation of risks before incidents occur
B. Simplification of audits
C. Reduction in documentation
D. Post-incident response
Q225. What is the purpose of a risk matrix in the risk assessment process?
A. To calculate audit frequency
B. To identify training needs
C. To evaluate and prioritize hazards based on likelihood and severity
D. To record near-miss events
Q226. Which type of training focuses on ensuring employees understand specific job hazards?
A. Task-specific safety training
B. Orientation training
C. General awareness training
D. Leadership development
Q227. What ensures that hazard controls remain effective over time?
A. Regular review and performance monitoring
B. One-time installation
C. Informal observation
D. Occasional policy revision
Q228. Which activity ensures the sustainability of continuous improvement efforts?
A. Limiting communication
B. Performing isolated audits
C. Reducing feedback channels
D. Integration of lessons learned into system updates
Q229. What is the role of top management in promoting safety culture?
A. Demonstrating leadership, providing resources, and setting expectations
B. Delegating all safety tasks
C. Limiting employee reporting

D. Conducting isolated reviews
Q230. Which safety management element validates that system components function as intended?
A. Safety Assurance
B. Safety Policy
C. Safety Promotion
D. Risk Assessment
Q231. Which activity enhances worker understanding of chemical hazards in the workplace?
A. Annual health checkups
B. Effective Hazard Communication (HAZCOM) training
C. Routine toolbox meetings
D. Maintenance scheduling
Q232. Which phase of the Plan-Do-Check-Act cycle involves implementing changes to correct deficiencies?
A. Act
B. Check
C. Do
D. Plan
Q233. What is the main benefit of establishing a clear chain of command in safety management?
A. Defined accountability and streamlined decision-making
B. Increased bureaucracy
C. Reduced communication
D. Informal authority
Q234. Which documentation process ensures traceability of safety improvements over time?
A. Verbal reporting
B. Informal file storage
C. Controlled document management system
D. Manual checklist only
Q235. What is the ultimate goal of a proactive safety management system?
A. To anticipate, identify, and mitigate risks before incidents occur
B. To respond reactively to accidents
C. To reduce training programs
D. To limit employee involvement
Q236. Which practice ensures continual effectiveness of a safety management system?
A. Regular internal auditing and review
B. Annual policy display
C. Informal meetings
D. Limited communication
Q237. What is the main advantage of near-miss reporting programs?
A. Simplifying audits
B. Reducing documentation
C. Limiting communication
D. Preventing accidents by addressing hazards before they cause harm
Q238. Which step in risk assessment involves determining potential impact of identified hazards?
A. Risk evaluation
B. Risk identification
C. Control verification
D. Incident analysis
Q239. Which safety management element ensures that the organization adapts to changes effectively?
A. Management of Change (MOC)

B. Risk avoidance
C. Administrative review
D. Annual planning

Q240. What is the primary benefit of job safety analysis (JSA)?
A. Reducing documentation
B. Assigning disciplinary actions
C. Identifying hazards and determining safe work procedures for each task
D. Tracking overtime

Q241. Which document communicates detailed procedures for managing workplace hazards?
A. Standard Operating Procedures (SOPs)
B. Risk register
C. Audit summary
D. Compliance certificate

Q242. What is the goal of implementing behavior-based safety programs?
A. To encourage safe actions and reduce at-risk behaviors through observation and feedback
B. To replace engineering controls
C. To simplify documentation
D. To reduce supervision

Q243. Which strategy best promotes continuous safety improvement?
A. Conducting annual reviews only
B. Implementing feedback loops that translate performance data into action
C. Minimizing communication
D. Limiting safety meetings

Q244. What ensures that safety training remains relevant and effective?
A. Regular evaluation and updates based on changing risks
B. Static training schedules
C. One-time orientation
D. Reduced refresher sessions

Q245. What is the primary objective of emergency drills?
A. To test readiness and improve emergency response effectiveness
B. To document employee attendance
C. To fulfill compliance only
D. To reduce communication

Q246. Which control method removes the hazard entirely from the workplace?
A. Administrative control
B. Substitution
C. Elimination
D. PPE

Q247. Which documentation provides a comprehensive record of identified risks and control measures?
A. Risk Register
B. Audit Log
C. Inspection Checklist
D. Incident Report

Q248. Which process ensures that lessons from incidents are integrated into system improvements?
A. Corrective and preventive action process
B. Routine monitoring
C. Documentation filing
D. Data storage

Q249. What ensures that risk communication reaches all affected personnel effectively?

A. Multichannel safety communication strategy
B. Single written notice
C. Restricted distribution
D. Informal conversation
Q250. Which safety principle emphasizes preventing exposure rather than responding to incidents?
A. Post-event review
B. Reactive investigation
C. Incident documentation
D. Proactive risk management
Q251. What is the key outcome of effective contractor safety management?
A. Consistent adherence to organizational safety standards by all contractors
B. Reduced operational oversight
C. Increased project speed
D. Reduced reporting
Q252. Which aspect of documentation control supports audit readiness?
A. Version tracking and timely updates
B. Paper-only storage
C. Verbal communication
D. Unrestricted document editing
Q253. Which process ensures that safety improvements remain measurable and results-oriented?
A. Establishing key performance indicators (KPIs)
B. Routine observations
C. Document archiving
D. Incident recording only
Q254. What is the purpose of conducting periodic risk reassessments?
A. To verify documentation format
B. To ensure that implemented controls remain effective amid changing conditions
C. To update staff lists
D. To satisfy policy requirements
Q255. Which method best verifies employee competence in performing safety-critical tasks?
A. Performance-based evaluation
B. Written policy distribution
C. Attendance tracking
D. Self-assessment form
Q256. Which activity best verifies the long-term success of safety programs?
A. Continuous performance monitoring and trend analysis
B. Annual awards ceremonies
C. Management delegation
D. Informal safety talks
Q257. What is the core purpose of a safety audit?
A. To reduce administrative work
B. To document employee attendance
C. To evaluate compliance and effectiveness of safety management processes
D. To verify payroll accuracy
Q258. Which type of control reduces human exposure by altering work practices or schedules?
A. Administrative control
B. Engineering control
C. Substitution
D. PPE
Q259. What is the purpose of a management system policy statement?

A. To declare organizational commitment to health and safety
B. To assign financial penalties
C. To define audit frequency
D. To summarize workforce numbers

Q260. What is the key benefit of cross-functional safety teams?

A. Limiting feedback
B. Reducing participation
C. Increasing bureaucracy
D. Combining diverse expertise to solve complex safety issues

Q261. Which element of safety communication ensures understanding among all employees?

A. Clarity, consistency, and accessibility of messages
B. Use of technical jargon
C. Limited distribution
D. One-time announcements

Q262. Which system ensures that all safety documents are updated and available to employees?

A. Document control system
B. Payroll management system
C. Procurement platform
D. Performance appraisal system

Q263. What is the primary goal of safety inspections?

A. To assign fault
B. To identify and correct hazardous conditions before incidents occur
C. To reduce management involvement
D. To archive data only

Q264. Which factor best demonstrates a mature safety culture?

A. Active employee engagement in identifying and mitigating risks
B. Reliance on management alone
C. Minimal reporting
D. Focus on lagging indicators only

Q265. What ensures that emergency response plans remain effective?

A. Regular review, updates, and drills
B. Annual printing
C. Distribution without feedback
D. Informal awareness

Q266. Which tool identifies root causes behind recurring incidents?

A. Root Cause Analysis (RCA)
B. Trend charting
C. Safety audit
D. Risk register

Q267. Which practice best supports continual hazard awareness among employees?

A. Online reminders only
B. Annual policy reviews
C. Frequent toolbox talks and safety briefings
D. Post-incident reviews only

Q268. Which management activity drives accountability across safety programs?

A. Assigning clear roles, responsibilities, and performance expectations
B. Delegating informally
C. Reducing reporting
D. Avoiding documentation

Q269. Which outcome is achieved through a proactive reporting culture?

A. Increased identification of hazards and preventive actions
B. Decreased engagement
C. Reduced compliance
D. Limited communication

Q270. What is the advantage of using leading indicators in safety performance measurement?

A. They delay response
B. They record historical data only
C. They measure compliance only
D. They provide early warnings that allow preventive action

Q271. Which activity confirms that all risk controls are functioning effectively?

A. Verification and validation testing
B. Document archiving
C. Employee surveys
D. Safety meetings

Q272. What is the main goal of an employee recognition program in safety management?

A. To penalize errors
B. To reinforce positive behaviors and encourage safe performance
C. To track attendance
D. To reduce feedback

Q273. What ensures safety procedures remain practical and relevant?

A. Periodic review and worker consultation
B. Strict management enforcement
C. Annual rewriting
D. Document storage

Q274. Which factor demonstrates leadership's true commitment to safety?

A. Consistent allocation of resources and visible engagement
B. Reactive decision-making
C. Delegating responsibilities without oversight
D. Occasional meetings

Q275. What is the primary function of safety performance benchmarking?

A. To meet basic compliance
B. To assess payroll accuracy
C. To compare safety results with industry standards and identify improvement opportunities
D. To manage procurement

Q276. Which principle best represents the goal of ergonomics?

A. Fit the person to the job
B. Fit the job to the person
C. Maintain average job design
D. Eliminate human variability

Q277. Which of the following is a primary ergonomic risk factor?

A. Task autonomy
B. Repetition
C. Motivation
D. Cognitive engagement

Q278. What does the NIOSH Lifting Equation primarily evaluate?

A. Static posture tolerance
B. Tool vibration exposure
C. Aerobic capacity
D. Safe lifting capacity

Q279. Which posture is most biomechanically efficient when lifting an object from the floor?

A. Twisting the spine
B. Bending forward at the waist
C. Using leg muscles while keeping the back straight
D. Reaching forward with arms extended

Q280. According to anthropometric design principles, what percentile is typically used to set minimum clearance dimensions?
A. 1st percentile female
B. 5th percentile female
C. 50th percentile male
D. 95th percentile male

Q281. Which metric best quantifies hand activity and force in repetitive work?
A. TLV for noise
B. RULA score
C. HAL (Hand Activity Level)
D. REBA score

Q282. Which factor does not directly contribute to ergonomic risk?
A. Force
B. Posture
C. Vibration
D. Incentive pay

Q283. What is the most critical goal of workstation design?
A. Minimize cost
B. Maximize aesthetic appeal
C. Maintain neutral body posture
D. Encourage high exertion levels

Q284. Which condition is most commonly associated with repetitive hand motions?
A. Carpal Tunnel Syndrome
B. Epicondylitis
C. Bursitis
D. Tendon rupture

Q285. When designing a seated workstation, the most ergonomic elbow angle for typing is approximately:
A. 45°
B. 60°
C. 90°
D. 120°

Q286. What is the purpose of the Rapid Upper Limb Assessment (RULA)?
A. Evaluate lifting limits
B. Assess hand-arm vibration
C. Evaluate postural load in upper body tasks
D. Measure aerobic workload

Q287. Which environmental factor most affects thermal comfort in office ergonomics?
A. Relative humidity
B. Noise level
C. Light intensity
D. Color scheme

Q288. The ideal viewing distance for a computer monitor is approximately:
A. 20 cm
B. 30 cm
C. 50–70 cm

D. 100 cm

Q289. Which control strategy is most effective for preventing cumulative trauma disorders?

A. Personal protective equipment
B. Engineering controls
C. Administrative controls
D. Worker selection

Q290. Which physical risk factor most contributes to low back disorders?

A. Awkward postures
B. Light repetitive motions
C. Prolonged sitting only
D. Cold environment

Q291. Which ergonomic principle applies when designing work for multiple operators of varying sizes?

A. Fixed workstation heights
B. Design for 50th percentile user
C. Use of adjustable features
D. Design for 95th percentile only

Q292. Which body area is most affected by poor seated posture?

A. Shoulders
B. Neck
C. Lower back
D. Knees

Q293. What is the main objective of anthropometric measurement in ergonomics?

A. Determine muscle endurance
B. Define workspace dimensions
C. Measure lighting intensity
D. Assess stress tolerance

Q294. Which ergonomic control prevents hand-arm vibration syndrome (HAVS)?

A. Administrative job rotation
B. Anti-vibration tool handles
C. Rest breaks every 2 hours
D. PPE such as gloves only

Q295. In task analysis, what is the main purpose of observing worker performance?

A. Identify noncompliance
B. Quantify physical load
C. Detect unsafe behaviors
D. Evaluate ergonomic risk

Q296. Which factor most influences visual strain in computer use?

A. Room temperature
B. Screen glare
C. Keyboard height
D. Chair lumbar support

Q297. Which measure best reduces neck strain during computer work?

A. Raising monitor above eye level
B. Aligning monitor top at eye level
C. Tilting head forward
D. Positioning monitor to the side

Q298. Which principle guides tool handle design for power grip tasks?

A. Small diameter to fit fingertips
B. Large diameter exceeding palm width

C. Diameter allowing full finger contact
D. Smooth metal surface only
Q299. What is the preferred control location for frequently used machine controls?
A. Above shoulder level
B. Below knee level
C. Within normal reach zone
D. Behind operator
Q300. Which administrative control reduces ergonomic strain from repetitive tasks?
A. Fixed work schedules
B. Job rotation and rest pauses
C. Task elimination
D. Increased work speed

Answer keys and Explanations (201-300)

Q201. Answer: A

Explanation: A Safety Manual outlines the structure, responsibilities, and operational processes that form the foundation of the organization's safety program. It serves as a reference for procedures, standards, and expectations, ensuring uniform application of safety principles. This document facilitates compliance, consistency, and continual improvement across all functional levels.

Q202. Answer: B

Explanation: A hazard identification program systematically detects potential sources of harm across processes, equipment, and behaviors. Early identification enables proactive control implementation, preventing accidents and minimizing exposure. This approach fosters a preventive culture where hazards are addressed before causing damage, strengthening overall safety system resilience.

Q203. Answer: A

Explanation: Safety KPIs provide quantifiable metrics to evaluate how well an organization achieves its safety goals. They enable ongoing monitoring, comparison, and improvement by translating data into actionable insights. Regular review of KPIs ensures that management decisions are data-driven and aligned with continuous safety improvement objectives.

Q204. Answer: A

Explanation: Integrating safety into business planning ensures that operational and strategic decisions consider safety implications. This approach embeds safety performance as a metric of organizational success. It aligns resource allocation, performance measurement, and leadership accountability with long-term health and safety priorities.

Q205. Answer: C

Explanation: Incident trend analysis examines accumulated data to identify repeating causes and patterns across events. This helps management address systemic weaknesses rather than isolated cases. By focusing on trends, organizations can implement targeted preventive measures that strengthen safety performance and reduce recurrence over time.

Q206. Answer: A

Explanation: Residual risk represents the amount of risk that persists even after applying all feasible mitigation measures. It acknowledges that total risk elimination is rarely possible. Understanding residual risk allows organizations to determine acceptability levels and decide whether additional controls or procedural safeguards are required.

Q207. Answer: A

Explanation: Gap analysis compares current safety performance against desired standards or regulatory benchmarks to reveal deficiencies. This diagnostic tool guides improvement plans by identifying areas requiring attention. Through structured comparison, gap analysis prioritizes corrective actions that close performance gaps and enhance system effectiveness.

Q208. Answer: D

Explanation: An action tracking and verification system documents corrective measures, assigns responsibility, and confirms completion. This structured approach ensures accountability and consistency, allowing management to monitor progress and effectiveness. Verified corrective actions prevent recurrence, demonstrating due diligence and commitment to continuous safety improvement.

Q209. Answer: A

Explanation: Root cause analysis investigates underlying system failures contributing to incidents. Sharing lessons learned ensures that identified weaknesses are corrected organization-wide. This open learning approach enhances transparency, prevents recurrence, and builds collective competence, reinforcing a proactive and adaptive safety culture.

Q210. Answer: A

Explanation: Safety culture assessments evaluate collective beliefs, attitudes, and behaviors related to safety. These insights help organizations understand the maturity and effectiveness of their culture. By identifying strengths and areas for improvement, assessments guide targeted interventions that promote trust, communication, and engagement, all of which are essential for sustainable safety success.

Q211. Answer: C

Explanation: Continuous monitoring and internal verification provide ongoing oversight of safety performance. This ensures that established controls remain effective and deviations are promptly corrected. Regular verification supports operational stability, compliance, and continuous improvement by maintaining vigilance between scheduled audit cycles.

Q212. Answer: A

Explanation: A non-punitive reporting policy assures employees that they can report hazards and errors without fear of punishment. This openness encourages early identification of issues and facilitates learning. Trust and transparency improve participation rates, allowing the organization to act proactively to prevent incidents and strengthen its safety management system.

Q213. Answer: A

Explanation: A management system maturity model assesses how advanced and integrated an organization's safety processes and culture are. It provides benchmarks for evaluating progress from basic compliance to proactive excellence. Using maturity models helps leaders prioritize initiatives and allocate resources strategically to achieve continuous growth in safety performance.

Q214. Answer: B

Explanation: Performance feedback translates monitoring results into actionable guidance for individuals and teams. It highlights strengths, identifies improvement areas, and reinforces learning. Constructive feedback fosters accountability and motivates consistent performance, turning data into progress and supporting a cycle of ongoing system refinement.

Q215. Answer: A

Explanation: Accurate, validated, and transparent data form the basis of credible safety analysis. Reliable data ensure that conclusions and decisions are well-founded, supporting evidence-based management. Transparency builds trust among stakeholders and strengthens the organization's ability to make effective, informed improvements to its safety management system.

Q216. Answer: C

Explanation: Safety data management and analysis ensure that all decisions are informed by verified, up-to-date information. This process involves systematic data collection, validation, and interpretation to identify trends and emerging risks. It transforms raw information into actionable intelligence that guides preventive strategies and resource allocation. Reliable data enables predictive insights, improves decision-making accuracy, and supports continuous improvement in organizational safety performance.

Q217. Answer: A

Explanation: Follow-up audits and performance evaluations confirm whether corrective actions achieve their intended outcomes. This verification ensures that identified deficiencies are truly resolved and that the implemented controls function effectively in practice. The process reinforces accountability, validates the reliability of safety systems, and promotes transparency. Regular verification safeguards against reoccurrence, ensuring that safety improvements are permanent and measurable across operations.

Q218. Answer: A

Explanation: Safety Promotion emphasizes education, awareness, and communication to create a culture of learning and improvement. It ensures that safety information, best practices, and lessons learned from incidents are shared across the organization. By encouraging open dialogue and continuous knowledge exchange, Safety Promotion strengthens engagement and cultivates collective responsibility for maintaining a safe and informed work environment.

Q219. Answer: D

Explanation: Safety performance evaluation measures how well an organization's safety initiatives meet established objectives. It involves analyzing data, reviewing incident trends, and assessing compliance with standards. These evaluations reveal opportunities for enhancement, ensuring that safety programs remain relevant and effective. The process drives informed decisions, accountability, and continuous improvement within the safety management framework.

Q220. Answer: A

Explanation: Employee participation programs empower workers to contribute directly to hazard identification and safety planning. This collaboration leverages frontline experience to develop practical, effective solutions. Active involvement enhances engagement, promotes ownership, and ensures that control measures are realistic. Such programs build trust, strengthen communication, and drive a safety culture grounded in shared responsibility and cooperation.

Q221. Answer: A

Explanation: Safety leadership is most effective when leaders model desired behaviors and demonstrate visible commitment to safety. Consistent engagement—such as participating in safety meetings, inspections, and training—reinforces expectations and builds credibility. Employees emulate leaders who actively uphold safety standards, fostering a culture of accountability, trust, and shared dedication to maintaining safe operations.

Q222. Answer: B

Explanation: Management review provides senior leadership with an opportunity to evaluate safety performance and system effectiveness. It ensures that objectives are met, emerging risks are addressed, and resources remain sufficient. This review strengthens accountability and alignment between operational performance and strategic direction, ensuring ongoing adaptation and improvement within the organization's safety management system.

Q223. Answer: A

Explanation: Engineering controls reduce exposure by physically modifying equipment, systems, or workspaces to isolate hazards. Examples include ventilation, machine guarding, and noise enclosures. These controls are highly effective because they act directly on the hazard source, minimizing dependence on human behavior. Implementing engineering controls ensures long-term risk reduction and promotes sustainable workplace safety.

Q224. Answer: A

Explanation: Proactive hazard reporting enables early recognition of unsafe conditions and behaviors, allowing corrective actions before harm occurs. Encouraging employees to report near misses and potential hazards strengthens prevention and cultivates awareness. This proactive culture enhances collaboration, transparency, and accountability while significantly reducing incident likelihood through continuous feedback and prompt intervention.

Q225. Answer: C

Explanation: A risk matrix visually ranks hazards by comparing their likelihood of occurrence with potential severity. This structured approach helps prioritize control actions, ensuring that high-risk issues receive immediate attention. It simplifies communication of complex risk information, guiding informed decision-making and efficient resource allocation to prevent serious incidents.

Q226. Answer: A

Explanation: Task-specific safety training provides workers with detailed instructions on the hazards and safe practices related to their particular duties. It addresses operational risks, proper equipment use, and emergency procedures. By tailoring content to specific job roles, this training enhances hazard recognition and control, ensuring that employees perform tasks safely and confidently.

Q227. Answer: A

Explanation: Regular review and performance monitoring ensure that hazard controls maintain their intended effectiveness as conditions evolve. This continuous evaluation detects wear, procedural drift, or emerging risks. By reassessing controls periodically, organizations sustain reliability, compliance, and protection levels, fostering a dynamic system of continuous improvement and resilience.

Q228. Answer: D

Explanation: Integrating lessons learned from incidents, audits, and observations ensures that continuous improvement becomes a living process. Each improvement cycle builds on practical experience, embedding updated knowledge into procedures and policies. This systematic incorporation transforms data and feedback into long-term organizational growth, reinforcing adaptability and commitment to excellence in safety performance.

Q229. Answer: A

Explanation: Top management drives safety culture by setting clear expectations, allocating resources, and modeling commitment through behavior. Leadership visibility reinforces priorities and empowers employees to engage. Their active involvement integrates safety into business goals, demonstrating accountability and shaping an organizational environment where safety excellence becomes a shared value.

Q230. Answer: A

Explanation: Safety Assurance validates the effectiveness of safety system components through monitoring, audits, and evaluations. It ensures implemented controls achieve desired results and identifies opportunities for enhancement. Continuous assurance fosters confidence, accountability, and reliability, ensuring that the safety management system adapts effectively to evolving risks and operational demands.

Q231. Answer: B

Explanation: Effective Hazard Communication (HAZCOM) training educates employees about hazardous chemicals, label interpretation, and Safety Data Sheet (SDS) use. It ensures workers understand potential exposures, protective measures, and emergency actions. Clear, consistent communication empowers safe handling, reinforces regulatory compliance, and cultivates awareness essential for chemical safety management.

Q232. Answer: A

Explanation: The "Act" phase implements corrective and preventive actions based on performance evaluations and audit findings. This stage ensures that identified deficiencies are resolved, and system improvements are embedded into operations. It completes the PDCA feedback loop, reinforcing continuous improvement and maintaining alignment between strategy and execution.

Q233. Answer: A

Explanation: A defined chain of command clarifies responsibilities and ensures efficient communication during routine operations or emergencies. It enables timely decision-making and consistent enforcement of safety policies. Clear hierarchy enhances accountability, coordination, and response capability, supporting disciplined, effective safety management practices throughout the organization.

Q234. Answer: C

Explanation: A controlled document management system organizes, tracks, and updates safety-related records systematically. It ensures traceability of revisions, preserves data integrity, and facilitates quick access to verified information. This structured documentation process supports audits, compliance verification, and continuous improvement by maintaining an accurate record of system evolution.

Q235. Answer: A

Explanation: A proactive safety management system aims to identify and control hazards before they result in harm. By emphasizing prevention, it uses data, analysis, and participation to anticipate risks and implement timely measures. This forward-thinking approach builds resilience, improves performance, and establishes a culture centered on continuous prevention and protection.

Q236. Answer: A

Explanation: Regular internal auditing and review provide objective assessment of system performance. These activities identify deficiencies, verify compliance, and ensure controls function as intended. Continuous auditing supports transparency, accountability, and improvement, allowing management to adapt to emerging risks. This proactive oversight maintains the system's reliability and effectiveness over time.

Q237. Answer: D

Explanation: Near-miss reporting allows organizations to capture valuable information about potential hazards before they lead to incidents. This proactive approach helps identify system weaknesses, improve controls, and enhance safety awareness. By treating near misses as learning opportunities, organizations strengthen prevention efforts, build engagement, and foster a proactive safety culture.

Q238. Answer: A

Explanation: Risk evaluation assesses how likely a hazard is to occur and the potential severity of its consequences. This process quantifies risk exposure, guiding prioritization of control efforts. By understanding potential impact, organizations can allocate resources effectively and prevent high-consequence events through focused mitigation strategies.

Q239. Answer: A

Explanation: Management of Change (MOC) ensures that operational, organizational, or process modifications are evaluated for safety implications before implementation. It maintains control over risk by analyzing potential impacts and ensuring appropriate safeguards are applied. This structured approach minimizes unforeseen hazards and supports a stable, adaptive safety system.

Q240. Answer: C

Explanation: Job Safety Analysis (JSA) systematically examines individual job steps to identify hazards and develop safe work practices. This structured review enhances awareness of specific risks and fosters preventive action. JSAs engage employees in safety ownership, improve training quality, and integrate risk control directly into daily operations.

Q241. Answer: A

Explanation: Standard Operating Procedures (SOPs) provide clear, step-by-step instructions for performing tasks safely. They standardize operations, reduce variability, and promote consistent adherence to safety standards. SOPs serve as a training resource and compliance reference, ensuring that all employees understand the correct methods for hazard management and control.

Q242. Answer: A

Explanation: Behavior-based safety programs focus on observing, analyzing, and reinforcing worker behaviors that affect safety. Through positive feedback and coaching, they reduce unsafe actions and promote consistent adherence to best practices. These programs empower employees to participate actively in prevention, creating a culture centered on accountability and continuous improvement.

Q243. Answer: B

Explanation: Feedback loops collect and analyze safety performance data, then convert insights into improvement measures. This process ensures that lessons learned drive change rather than remaining theoretical. Integrating feedback into policies and procedures fosters adaptability, accountability, and consistent advancement of the organization's safety objectives.

Q244. Answer: A

Explanation: Regularly reviewing and updating safety training ensures that content reflects new hazards, regulations, and technologies. This adaptability keeps employees informed and competent in current procedures. Ongoing evaluation aligns training with real-world needs, reinforcing skills and ensuring continual readiness in a dynamic work environment.

Q245. Answer: A

Explanation: Emergency drills simulate crisis scenarios to assess preparedness and response efficiency. These exercises reveal weaknesses, validate plans, and train employees to act swiftly and correctly under pressure. Regular drills enhance confidence, coordination, and competency, ensuring effective responses that protect personnel and assets during real emergencies.

Q246. Answer: C

Explanation: Elimination completely removes the hazard source, preventing any exposure risk. It is the most effective control method within the hierarchy of controls. Eliminating hazards may involve redesigning processes, changing materials, or discontinuing dangerous tasks, providing the highest level of protection for workers and operations.

Q247. Answer: A

Explanation: A risk register compiles all identified hazards, assessed risk levels, and applied control measures in one centralized document. It facilitates tracking, accountability, and prioritization of risk management activities. The register ensures visibility across departments, promoting consistency and informed decision-making for ongoing safety management.

Q248. Answer: A

Explanation: Corrective and preventive action processes convert incident findings into actionable improvements. They address root causes, implement long-term solutions, and prevent recurrence. Integrating these actions into policies and systems strengthens reliability, promotes learning, and drives continuous enhancement of safety performance across the organization.

Q249. Answer: A

Explanation: A multichannel safety communication strategy uses meetings, alerts, digital platforms, and signage to reach every employee effectively. Diverse communication channels ensure accessibility, comprehension, and timely dissemination of information. This inclusive approach enhances awareness, reinforces accountability, and builds a shared understanding of safety expectations across the organization.

Q250. Answer: D

Explanation: Proactive risk management focuses on anticipating and eliminating hazards before harm occurs. It relies on data analysis, predictive indicators, and employee engagement to identify potential issues early. This forward-looking approach reduces incident frequency and fosters a culture committed to continuous prevention and long-term operational resilience.

Q251. Answer: A

Explanation: Effective contractor safety management ensures that external personnel follow the same safety standards as internal employees. It establishes expectations through training, orientation, and performance monitoring. Consistency across all parties reduces risk, enhances collaboration, and promotes a unified safety culture that maintains high operational integrity.

Q252. Answer: A

Explanation: Version tracking and timely updates ensure that safety documents remain accurate, current, and compliant with standards. This traceability simplifies audits by providing verifiable records of revisions and approvals. Controlled documentation management reinforces accountability, reliability, and transparency in safety communication and compliance reporting.

Q253. Answer: A

Explanation: Key performance indicators (KPIs) quantify safety performance in terms of goals, trends, and compliance. They provide management with measurable benchmarks for evaluating the success of safety initiatives. Tracking KPIs ensures continuous improvement remains structured, data-driven, and aligned with strategic organizational objectives.

Q254. Answer: B

Explanation: Periodic risk reassessment evaluates whether controls continue to mitigate risks effectively as operations, personnel, or equipment evolve. This proactive step identifies emerging hazards and ensures adaptability. Regular reassessments maintain accuracy of risk prioritization and support a dynamic, responsive safety management system.

Q255. Answer: A

Explanation: Performance-based evaluation directly observes employees completing safety-critical tasks, confirming both knowledge and practical skill. This hands-on approach identifies training gaps, reinforces safe behavior, and validates compliance with operational standards. It ensures workforce readiness and competence, maintaining reliability in high-risk environments.

Q256. Answer: A

Explanation: Continuous performance monitoring and trend analysis track long-term effectiveness by identifying improvements or declines in safety outcomes. This ongoing evaluation highlights patterns, validates interventions, and ensures sustained focus on key risks. Regular analysis enables evidence-based decision-making, fostering continuous improvement and ensuring the organization's safety programs remain adaptive and resilient over time.

Q257. Answer: C

Explanation: Safety audits systematically review programs, practices, and documentation to assess whether the safety management system meets internal and regulatory standards. By identifying strengths and areas for improvement, audits provide actionable insights for enhancing effectiveness.

This verification process builds accountability, ensures compliance, and reinforces a culture of continuous improvement and transparency.

Q258. Answer: A

Explanation: Administrative controls reduce risk through changes in work procedures, schedules, or policies to limit exposure time and enhance safe behavior. Examples include rotating tasks, setting safe work limits, and implementing training programs. These controls rely on supervision and employee compliance, supporting engineering and elimination strategies to achieve comprehensive hazard reduction.

Q259. Answer: A

Explanation: A management system policy statement expresses leadership's formal commitment to ensuring safe, healthy, and compliant operations. It provides direction and expectations for all personnel while reinforcing accountability at every level. This foundational document establishes the organization's dedication to continuous improvement and demonstrates visible top-level support for safety initiatives.

Q260. Answer: D

Explanation: Cross-functional safety teams integrate insights from different departments to identify, assess, and resolve safety challenges effectively. This collaboration fosters innovative problem-solving and ensures that varied operational perspectives inform decision-making. By leveraging collective expertise, these teams enhance communication, accountability, and the practical implementation of robust safety measures across the organization.

Q261. Answer: A

Explanation: Effective safety communication depends on clear, consistent, and accessible messaging tailored to all literacy and language levels. This approach ensures that critical information is understood and applied correctly. By promoting transparency and comprehension, organizations enhance compliance, engagement, and shared accountability in achieving safety goals.

Q262. Answer: A

Explanation: A document control system maintains organized, current, and accessible safety documentation. It tracks revisions, approvals, and distribution to ensure accuracy and compliance. Proper document control prevents confusion, supports audits, and guarantees that employees always have access to the latest policies, procedures, and training materials essential for safe operations.

Q263. Answer: B

Explanation: Safety inspections proactively detect unsafe conditions, behaviors, and equipment deficiencies that could lead to incidents. These routine checks ensure that corrective actions are implemented promptly. By maintaining vigilance, inspections reinforce accountability, promote hazard control, and sustain a culture of continuous prevention and improvement.

Q264. Answer: A

Explanation: A mature safety culture is characterized by employees who actively engage in hazard reporting, risk assessments, and improvement initiatives. This shared ownership creates a collaborative, transparent environment where everyone contributes to safety excellence. High engagement drives accountability, continuous learning, and long-term organizational resilience.

Q265. Answer: A

Explanation: Regularly reviewing, updating, and conducting drills validates the accuracy and practicality of emergency response plans. These exercises reveal weaknesses, test coordination, and enhance preparedness. Continuous refinement ensures that the organization can respond efficiently, minimizing harm and maintaining operational continuity during emergencies.

Q266. Answer: A

Explanation: Root Cause Analysis (RCA) investigates underlying system, process, or human factors contributing to repeated incidents. It focuses on eliminating causes rather than symptoms. RCA supports sustainable improvements, enabling organizations to strengthen controls, prevent recurrence, and advance long-term safety performance through targeted corrective action.

Q267. Answer: C

Explanation: Frequent toolbox talks and safety briefings maintain active awareness of hazards and safe practices. These sessions promote open dialogue, encourage feedback, and reinforce training. Regular engagement keeps safety knowledge current and practical, building consistency and vigilance across all levels of the workforce.

Q268. Answer: A

Explanation: Assigning defined roles and measurable expectations ensures that every employee understands their safety responsibilities. This structured accountability system fosters ownership, clarity, and consistent follow-through. Clearly articulated duties align team efforts with organizational objectives, enhancing transparency and the overall effectiveness of safety management.

Q269. Answer: A

Explanation: A proactive reporting culture encourages employees to report near misses, unsafe conditions, and potential risks promptly. This open exchange enhances hazard visibility, enabling early intervention and prevention. By valuing input without punishment, organizations create trust, strengthen responsiveness, and sustain continuous safety improvement.

Q270. Answer: D

Explanation: Leading indicators track proactive measures such as training completion, inspections, and near-miss reports. These metrics reveal performance trends before incidents occur, enabling early correction. By focusing on prevention, leading indicators improve responsiveness, promote accountability, and drive continuous safety enhancement across operations.

Q271. Answer: A

Explanation: Verification and validation testing evaluate whether control measures perform as designed under operational conditions. This ensures that implemented safeguards reliably mitigate identified hazards. Continuous verification builds confidence in system integrity and prevents degradation of safety performance over time.

Q272. Answer: B

Explanation: Employee recognition programs motivate individuals and teams by rewarding safe conduct and proactive participation. Acknowledging positive behavior builds morale, promotes consistency, and embeds safety values into daily operations. Recognition reinforces desired outcomes, sustaining engagement and commitment to a strong safety culture.

Q273. Answer: A

Explanation: Periodic review and worker consultation ensure that safety procedures align with current operations and technologies. Engaging employees provides real-world feedback on applicability and effectiveness. This collaboration maintains accuracy, promotes ownership, and sustains continuous improvement across safety processes and documentation.

Q274. Answer: A

Explanation: Leadership commitment is proven through consistent resource allocation, active participation, and clear communication of priorities. Visible involvement shows employees that safety is integral to organizational success. This dedication builds trust, encourages participation, and solidifies a culture centered on prevention and excellence.

Q275. Answer: C

Explanation: Safety performance benchmarking compares organizational metrics against peers or recognized standards. This analysis reveals strengths, gaps, and best practices that drive progress. By setting measurable targets, benchmarking enhances competitiveness, fosters accountability, and inspires continuous advancement toward excellence in safety management performance.

Q276. Answer: B

Explanation: The fundamental ergonomic principle is to fit the job to the person, not the other way around. This ensures that work tasks, equipment, and environments accommodate the worker's physical and cognitive abilities. Designing for human variability minimizes fatigue, enhances comfort, and improves performance, reducing injury risk and promoting long-term efficiency in diverse work populations.

Q277. Answer: B

Explanation: Repetition is a primary ergonomic risk factor contributing to musculoskeletal disorders (MSDs). Tasks requiring repeated movements over long periods without rest strain muscles and tendons, especially in the hands, wrists, and shoulders. Proper job rotation, micro-breaks, and automation reduce repetitive stress, aligning with NIOSH and OSHA ergonomic guidelines for sustainable workload management.

Q278. Answer: D

Explanation: The NIOSH Lifting Equation assesses safe lifting capacity by calculating a Recommended Weight Limit (RWL) based on six task variables, including load distance, frequency, and coupling quality. It determines the Lifting Index (LI), which quantifies ergonomic risk for manual material handling. The equation helps identify unsafe lifting conditions and guide redesign to reduce back and shoulder strain.

Q279. Answer: C

Explanation: Lifting with the legs while maintaining a straight back minimizes spinal compression and shear forces on the lumbar region. The large muscles of the legs and hips handle the majority of the load, reducing strain on the lower back. Proper lifting mechanics are central to ergonomic interventions in manual material handling tasks.

Q280. Answer: D

Explanation: Clearance dimensions, such as doorway heights or aisle widths, are usually designed for the 95th percentile male to ensure nearly all users can pass or fit comfortably. Designing for larger body sizes accommodates most of the population and avoids physical constraints that could lead to awkward postures, collisions, or restricted movement in the workspace.

Q281. Answer: C

Explanation: The Hand Activity Level (HAL) metric combines repetition rate and exerted force to evaluate risk for hand and wrist disorders. HAL, developed by ACGIH, helps determine whether task demands exceed the threshold limit value (TLV) for musculoskeletal risk. This metric guides ergonomic redesign of repetitive manual operations such as assembly or packaging.

Q282. Answer: D

Explanation: Ergonomic risk arises from physical factors such as force, posture, repetition, and vibration, not from motivational systems like incentive pay. While incentives can influence behavior and pace, they do not inherently cause biomechanical strain. Proper task design and work-rest balance are more effective in reducing ergonomic injury risks.

Q283. Answer: C

Explanation: Neutral posture minimizes stress on muscles and joints, allowing body segments to align naturally. Workstations designed for neutral posture prevent awkward positions that contribute to musculoskeletal disorders. Adjustable chairs, monitor heights, and reach zones support ergonomic adaptation, ensuring comfort and efficiency for users with different body dimensions.

Q284. Answer: A

Explanation: Carpal Tunnel Syndrome (CTS) results from compression of the median nerve in the wrist, often due to repetitive hand and wrist movements, sustained flexion, or poor workstation design. Ergonomic interventions such as wrist-neutral keyboard placement, reduced force, and micro-breaks effectively prevent CTS in office and assembly environments.

Q285. Answer: C

Explanation: A 90° elbow angle keeps the forearms parallel to the floor, allowing neutral shoulder and wrist positions. This reduces strain on the upper limbs and prevents tendon loading. Proper seating height and adjustable keyboard trays support optimal typing posture, minimizing long-term musculoskeletal stress.

Q286. Answer: C

Explanation: RULA assesses the risk of upper limb musculoskeletal disorders by scoring postures of the neck, trunk, and upper limbs. It identifies high-risk body positions requiring ergonomic correction. The method is widely used for office, manufacturing, and healthcare settings to guide workstation improvements and job redesign for better posture and movement efficiency.

Q287. Answer: A

Explanation: Relative humidity affects the body's ability to dissipate heat through evaporation. Maintaining indoor humidity between 40–60% enhances comfort and reduces fatigue or dehydration. Thermal comfort integrates humidity, air temperature, radiant heat, and air movement, making it a key parameter in ergonomic workspace climate control design.

Q288. Answer: C

Explanation: An optimal viewing distance of 50–70 cm (20–28 inches) allows clear focus with minimal eye strain. This distance reduces convergence effort and prevents leaning forward, which stresses the neck and shoulders. Combined with a top-of-screen-at-eye-level position, it ensures balanced visual ergonomics and postural alignment.

Q289. Answer: B

Explanation: Engineering controls, such as redesigning tools or workstations to fit human capabilities, eliminate or reduce ergonomic hazards at the source. They are more effective than administrative or PPE measures because they permanently alter task demands. Examples include adjustable platforms, mechanical assists, and improved handle designs.

Q290. Answer: A

Explanation: Awkward postures, especially trunk flexion and rotation under load, significantly increase spinal compression and shear forces. Repeated exposure leads to lower back disorders. Ergonomic redesign should promote neutral alignment, adjustable work heights, and mechanical aids to reduce bending and twisting motions during material handling.

Q291. Answer: C

Explanation: Adjustable workstations accommodate a wide range of body sizes and anthropometric variability. This flexibility ensures each user can maintain neutral posture and reach comfort zones. Designing adjustability into chairs, desks, and controls improves inclusivity, reduces fatigue, and supports productivity across diverse operator populations.

Q292. Answer: C

Explanation: Poor seated posture increases lumbar spine compression, leading to lower back pain and potential disc degeneration. Maintaining lumbar support and slight recline reduces spinal loading. Ergonomically designed chairs and appropriate monitor height promote correct alignment, minimizing lower back discomfort and long-term musculoskeletal injury risk.

Q293. Answer: B

Explanation: Anthropometric data define appropriate workspace dimensions by accounting for human body measurements and reach zones. Designers use these data to set equipment placement, seat height, and control panel spacing. Proper anthropometric design minimizes awkward posture and ensures operational comfort for users of varying body sizes.

Q294. Answer: B

Explanation: Anti-vibration handles and dampening technology reduce the transmission of vibration from power tools to the user's hands and arms. Engineering controls like these are the most effective means of preventing HAVS. Administrative or PPE solutions provide limited protection, while design modifications address the root mechanical cause.

Q295. Answer: D

Explanation: Observing worker performance helps identify postural stress, repetitive motion, and force application that contribute to ergonomic risk. This qualitative and quantitative assessment supports targeted redesign of tasks and tools. It provides insight into actual versus intended work methods, a core component of participatory ergonomic evaluation.

Q296. Answer: B

Explanation: Screen glare causes eye fatigue and headaches due to excessive luminance contrast. Positioning monitors perpendicular to light sources and using anti-glare screens improve visual ergonomics. Maintaining proper brightness and contrast settings reduces visual strain, supporting prolonged comfortable computer work and reducing risk of eye discomfort.

Q297. Answer: B

Explanation: Aligning the monitor top at eye level allows natural gaze angle and minimizes neck flexion. This posture keeps the cervical spine in neutral alignment, reducing neck and shoulder strain. Proper screen height and distance support visual comfort and prevent cumulative neck discomfort during extended computer use.

Q298. Answer: C

Explanation: Tool handles designed for full finger contact distribute force evenly across the palm, reducing localized pressure. An optimal diameter (approximately 30–50 mm) enhances power grip strength and comfort. Ergonomic handle materials and shape minimize slippage and muscle strain, improving tool control and safety.

Q299. Answer: C

Explanation: Frequent-use controls should be located within the operator's normal reach zone, typically between shoulder and waist height and within comfortable arm's reach. This placement minimizes overextension, improves reaction time, and reduces repetitive reaching fatigue. Applying reach envelope principles enhances efficiency and prevents awkward postures.

Q300. Answer: B

Explanation: Job rotation and scheduled rest pauses reduce repetitive stress by alternating muscle group use and allowing recovery periods. Administrative controls supplement engineering interventions to manage exposure duration. Implementing structured breaks helps maintain productivity while lowering the risk of repetitive strain injuries in repetitive or static work environments.

Questions (301-400)

Q301. Which ergonomic control most effectively reduces shoulder strain during overhead work?

A. Administrative rotation
B. Use of adjustable-height platforms
C. Training on lifting
D. Providing shoulder supports

Q302. Which of the following is an example of a neutral wrist posture?

A. Straight and aligned with forearm
B. Bent upward
C. Flexed downward
D. Twisted sideways

Q303. What is the recommended hand–arm vibration exposure limit according to ISO and ACGIH standards?

A. 1.5 m/s^2
B. 2.5 m/s^2
C. 5.0 m/s^2
D. 10 m/s^2

Q304. Which type of muscle contraction is most fatiguing during static work?

A. Concentric
B. Eccentric
C. Isometric
D. Dynamic

Q305. Which of the following best defines the term "ergonomic reach envelope"?

A. Area reachable by bending
B. Range of motion for joints
C. Space comfortably reachable without leaning
D. Maximum physical reach distance

Q306. Which design approach ensures most users can safely and comfortably use equipment?

A. Designing for the average person
B. Designing for extremes and adjustability
C. Designing for men only
D. Designing for efficiency over safety

Q307. Which work factor most contributes to cumulative trauma disorders (CTDs)?

A. High visual demand
B. High repetition and force
C. Noise exposure
D. Short duration tasks

Q308. Which control measure best reduces neck flexion for microscope users?

A. Adjustable chair height
B. Padded chin rest
C. Frequent breaks
D. Tilting eyepiece or angled microscope stand

Q309. Which tool design minimizes wrist deviation during repetitive screwing tasks?

A. Straight-handle screwdriver
B. Pistol-grip screwdriver
C. Thin metal handle
D. Long unbalanced handle

Q310. Which factor is key in determining ergonomic lifting frequency limits?

A. Gender of worker

B. Load coupling quality
C. Lifting hand dominance
D. Worker age

Q311. Which concept best supports ergonomic seating design?
A. Fixed backrest angle
B. Upright 90° backrest only
C. Slight recline of 100–110°
D. Forward tilt of 20°

Q312. What is the purpose of anthropometric data in control panel design?
A. To reduce visual strain
B. To determine control and display placement
C. To improve electrical efficiency
D. To standardize labeling

Q313. What is the optimal working height for tasks requiring fine manual precision?
A. 10 cm below elbow height
B. At elbow height
C. 5–10 cm above elbow height
D. 20 cm above shoulder height

Q314. Which factor has the greatest influence on human error probability in control operations?
A. Age of operator
B. Interface complexity
C. Shift timing
D. Tool material

Q315. Which task is most likely to cause shoulder impingement?
A. Pushing carts at waist level
B. Lifting objects above shoulder height
C. Reaching forward at desk height
D. Using a foot pedal repeatedly

Q316. Which design principle minimizes error in visual displays?
A. Color redundancy and grouping
B. Complex symbology
C. Random layout
D. Bright flashing colors

Q317. Which workstation feature reduces fatigue in prolonged standing tasks?
A. Hard flooring
B. Anti-fatigue mats
C. Fixed standing height
D. Narrow base support

Q318. Which control minimizes repetitive strain in data entry jobs?
A. Screen magnification
B. Split keyboard design
C. Low monitor brightness
D. Fixed seating height

Q319. Which lighting condition best supports visual ergonomics for computer use?
A. Direct overhead lighting
B. Colored decorative lighting
C. Bright focused spotlights
D. Diffuse ambient lighting

Q320. Which aspect of the NIOSH Lifting Equation accounts for frequency of lifts?
A. Load constant

B. Frequency multiplier
C. Coupling multiplier
D. Asymmetry multiplier

Q321. What is the primary focus of cognitive ergonomics?
A. Posture optimization
B. Decision-making and information processing
C. Force reduction
D. Anthropometric fitting

Q322. Which human factor most contributes to operator vigilance decline during monotonous tasks?
A. Environmental temperature
B. Motivation
C. Task variability
D. Lack of stimulation

Q323. What design element minimizes error in auditory warning systems?
A. High-frequency pure tones
B. Contextually distinct sounds
C. Continuous background beeps
D. Random volume changes

Q324. Which anthropometric value defines seated eye height?
A. Vertical reach
B. Elbow height
C. Eye level from seat reference point
D. Shoulder height

Q325. Which ergonomic design adjustment most effectively reduces forearm pronation during typing?
A. Palm rest use
B. Negative keyboard tilt
C. Elevated monitor position
D. Padded chair arms

Q326. Which ergonomic factor most influences the design of tool handles for repetitive work?
A. Tool weight
B. Handle surface texture
C. Grip span
D. Handle color

Q327. What ergonomic principle is emphasized when designing manual material handling tasks?
A. Lifting with one hand
B. Minimizing horizontal reach distance
C. Maximizing load height
D. Encouraging frequent bending

Q328. What is the main purpose of an ergonomic job hazard analysis (JHA)?
A. To assess financial losses
B. To evaluate ergonomic risk factors
C. To monitor employee attendance
D. To calculate exposure time for chemicals

Q329. Which measurement is most useful for setting workbench height?
A. Elbow height
B. Shoulder width
C. Sitting eye height
D. Knee height

Q330. Which ergonomic principle supports the design of adjustable seating?
A. Designing for extremes only

B. Designing for average body size
C. Designing for user adjustability
D. Designing for gender-specific differences

Q331. Which ergonomic principle applies to workstation layout to reduce unnecessary movement?
A. Arrange tools based on frequency of use
B. Randomly position components
C. Store heavy items on upper shelves
D. Place controls behind the operator

Q332. What is the primary advantage of sit-stand workstations?
A. Increased work speed
B. Improved posture variation
C. Higher energy consumption
D. Reduced task accuracy

Q333. Which principle minimizes ergonomic risk during tool use?
A. Increase grip force for better control
B. Maintain neutral wrist alignment
C. Keep wrist fully flexed
D. Use maximum muscle exertion

Q334. Which principle supports the effective design of visual displays in control rooms?
A. Use of high contrast and minimal clutter
B. Random display arrangement
C. Overlapping graphical elements
D. Minimal brightness variation

Q335. Which factor most affects lower back compression during lifting?
A. Vertical distance of lift
B. Weight of the load only
C. Worker footwear type
D. Temperature of the environment

Q336. What ergonomic control reduces eye strain in digital work environments?
A. Increased monitor brightness
B. Proper lighting contrast and screen position
C. Smaller font size
D. Blue background lighting

Q337. Which method effectively identifies high-risk postures during manual tasks?
A. Visual estimation
B. RULA or REBA assessment
C. Worker self-report only
D. Observation of productivity rates

Q338. Which design principle enhances hand-tool control accuracy?
A. Increasing tool mass
B. Removing finger grooves
C. Minimizing handle diameter
D. Using frictional handle materials

Q339. What is the purpose of ergonomic training programs?
A. Reinforce personal responsibility only
B. Teach proper body mechanics and awareness
C. Replace engineering solutions
D. Eliminate workstation variability

Q340. Which factor should be prioritized in designing tools to minimize forearm strain?
A. Handle material hardness

B. Balance and weight distribution
C. Color contrast of tool
D. Tool length only

Q341. Which concept is applied when designing multi-user control panels?
A. Adjustability and modularity
B. Fixed seat and reach settings
C. Compact layout without spacing
D. Equal dimensions for all controls

Q342. Which biomechanical principle explains reduced muscle force when joints deviate from neutral?
A. Leverage inefficiency
B. Muscle fatigue threshold
C. Elastic recovery
D. Force equilibrium

Q343. Which control measure prevents fatigue in standing workstations?
A. Fixed posture maintenance
B. Providing adjustable footrests
C. Narrow work surfaces
D. Limiting movement

Q344. Which ergonomic intervention reduces forward head posture in office work?
A. Lowering monitor height
B. Raising monitor to eye level
C. Tilting keyboard downward
D. Increasing chair recline

Q345. Which human factor most influences performance under time pressure?
A. Emotional intelligence
B. Anthropometric fit
C. Stress and cognitive load
D. Postural endurance

Q346. Which measure best reduces vibration exposure from power tools?
A. Wearing thick gloves
B. Using anti-vibration mounts and dampers
C. Taking long lunch breaks
D. Applying extra grip pressure

Q347. Which factor is most important in preventing musculoskeletal injury from prolonged sitting?
A. High seat cushion stiffness
B. Proper lumbar support
C. Armrest removal
D. Fixed back angle

Q348. What does the term "biomechanical efficiency" refer to in ergonomics?
A. Maximum muscle size
B. Maximum energy consumption
C. Reduced rest periods
D. Optimal joint leverage and force application

Q349. Which factor improves visibility and accuracy in precision assembly work?
A. High ambient lighting contrast
B. Adjustable localized task lighting
C. Reduced illumination levels
D. Colored backlighting

Q350. Which ergonomic approach best integrates human factors into process safety?

A. Designing systems with user-centered interfaces
B. Relying on worker adaptation
C. Prioritizing equipment cost over usability
D. Limiting operator feedback mechanisms

Q351. Which element of the fire tetrahedron represents the chemical reaction necessary for combustion?
A. Heat
B. Oxygen
C. Fuel
D. Chain reaction

Q352. Which fire class involves energized electrical equipment?
A. Class A
B. Class B
C. Class C
D. Class D

Q353. Which NFPA standard governs the use and maintenance of portable fire extinguishers?
A. NFPA 10
B. NFPA 13
C. NFPA 25
D. NFPA 30

Q354. What is the minimum flash point for a liquid classified as flammable under NFPA 30?
A. Below 100°F (37.8°C)
B. 100°F to 140°F (37.8–60°C)
C. Above 140°F (60°C)
D. Below 200°F (93°C)

Q355. Which type of sprinkler system uses air pressure instead of water in its piping until activation?
A. Wet pipe system
B. Dry pipe system
C. Deluge system
D. Preaction system

Q356. Which principle underlies the use of carbon dioxide extinguishers?
A. Cooling by water absorption
B. Smothering by oxygen displacement
C. Inhibiting chain reaction
D. Fuel separation

Q357. Which color identifies water-based fire extinguishers according to NFPA standards?
A. Red
B. Green
C. Silver
D. Yellow

Q358. Which condition best describes a Class D fire?
A. Involves metals like magnesium and titanium
B. Involves cooking oils and fats
C. Involves flammable gases
D. Involves energized electrical equipment

Q359. What is the primary hazard of combustible dust?
A. Explosion potential when suspended in air
B. Corrosion of metal surfaces
C. Fire spread by conduction
D. Moisture absorption

Q360. Which fire detection device responds to changes in light obscuration caused by smoke particles?
A. Photoelectric detector
B. Ionization detector
C. Heat detector
D. Flame detector
Q361. Which NFPA code provides requirements for the installation of sprinkler systems?
A. NFPA 72
B. NFPA 25
C. NFPA 13
D. NFPA 70E
Q362. Which extinguisher is suitable for use on Class K fires?
A. Wet chemical extinguisher
B. CO_2 extinguisher
C. Water extinguisher
D. Halon extinguisher
Q363. Which statement describes the purpose of a fire wall?
A. To limit fire spread between building sections
B. To act as a smoke detection system
C. To provide ventilation
D. To cool surfaces
Q364. What is the first step in the fire risk assessment process?
A. Estimating loss potential
B. Identifying potential fire hazards
C. Installing alarms
D. Reviewing insurance coverage
Q365. Which control measure best prevents ignition of flammable vapors?
A. Controlling static electricity
B. Increasing humidity
C. Using water sprays
D. Isolating fuel containers
Q366. Which OSHA regulation addresses fire protection in general industry?
A. 29 CFR 1910 Subpart L
B. 29 CFR 1910 Subpart Z
C. 29 CFR 1926 Subpart P
D. 29 CFR 1910 Subpart D
Q367. Which device detects flame radiation in industrial areas?
A. Thermal detector
B. Photoelectric detector
C. Ultraviolet/Infrared (UV/IR) detector
D. Smoke aspirating system
Q368. Which fire suppression system uses inert gases to reduce oxygen concentration?
A. Clean agent system
B. Foam system
C. Wet pipe system
D. Water mist system
Q369. What is the most effective method for preventing spontaneous combustion of oily rags?
A. Storing in self-closing metal containers
B. Leaving them to dry in open air
C. Washing with water only

D. Stacking in piles

Q370. Which type of sprinkler activates upon detection of heat at the sprinkler head?

A. Wet pipe

B. Dry pipe

C. Preaction

D. Deluge

Q371. Which flammable liquid storage cabinet color is typically used according to OSHA and NFPA?

A. Green

B. Red

C. Blue

D. Yellow

Q372. Which type of foam is most effective for hydrocarbon fuel fires?

A. Aqueous Film Forming Foam (AFFF)

B. Protein foam

C. Alcohol-resistant foam

D. High-expansion foam

Q373. Which fire detection device is best for clean rooms or data centers?

A. Air sampling smoke detector

B. Ionization detector

C. Heat detector

D. Manual pull station

Q374. What is the primary purpose of fire separation distance for flammable storage tanks?

A. To enhance ventilation

B. To improve accessibility for maintenance

C. To reduce radiant heat exposure and prevent fire spread

D. To optimize space utilization

Q375. Which fire suppression agent is environmentally friendly and leaves no residue?

A. FM-200 (HFC-227ea)

B. Halon 1211

C. Dry chemical powder

D. Carbon dioxide

Q376. What is the primary purpose of a preaction sprinkler system?

A. To prevent accidental discharge in sensitive areas

B. To deliver higher water pressure

C. To cool combustible materials continuously

D. To reduce installation costs

Q377. Which NFPA standard governs fire alarm systems?

A. NFPA 72

B. NFPA 25

C. NFPA 10

D. NFPA 101

Q378. What hazard control method is most effective for preventing ignition in flammable gas storage?

A. Use of dry chemical extinguishers

B. Elimination of ignition sources

C. Pressure relief devices

D. Flame arrestors only

Q379. Which component ensures effective ventilation in flammable liquid storage rooms?

A. Explosion-proof exhaust system

B. Open window

C. Ceiling fan

D. Air-conditioning unit

Q380. Which class of fire extinguishers should be used on cooking oil fires?

A. Class K

B. Class B

C. Class D

D. Class A

Q381. What is the key advantage of foam-based fire suppression systems?

A. Compatibility with electrical fires

B. High pressure water discharge

C. Ability to smother flammable liquid fires by forming a vapor-sealing blanket

D. Rapid heat absorption only

Q382. Which property defines a liquid's ease of ignition?

A. Flash point

B. Boiling point

C. Autoignition temperature

D. Density

Q383. What is the main purpose of fire drills in the workplace?

A. To measure smoke levels

B. To verify alarm sound levels

C. To assess system voltage

D. To test and improve employee readiness during emergencies

Q384. Which type of extinguisher is appropriate for flammable liquid fires?

A. Carbon dioxide extinguisher

B. Water extinguisher

C. Wet chemical extinguisher

D. Class D powder extinguisher

Q385. What is the function of an intumescent coating in fire protection?

A. To reduce static buildup

B. To expand under heat and insulate structural steel

C. To reflect radiant heat

D. To absorb sound waves

Q386. Which gas is commonly used in clean agent systems for data centers?

A. FM-200 (HFC-227ea)

B. CO_2

C. Halon 1301

D. Argon

Q387. Which method minimizes the spread of combustible dust explosions?

A. Proper containment and explosion venting

B. Use of water sprinklers only

C. Installing larger ducts

D. Increasing air pressure

Q388. What does the fire load of a building represent?

A. Water flow rate available

B. Amount of electrical wiring installed

C. Total potential heat energy released by combustible materials

D. Occupant density

Q389. Which device automatically closes to contain fire and smoke spread through ventilation ducts?

A. Fire damper

B. Check valve

C. Fan guard

D. Pressure regulator

Q390. Which factor primarily influences the design of fire suppression water supply systems?

A. Available pressure and flow rate

B. Color of hydrants

C. Building aesthetics

D. Distance from main entrance

Q391. What is the purpose of grounding and bonding in flammable liquid handling areas?

A. To detect leaks automatically

B. To improve lighting efficiency

C. To stabilize equipment vibration

D. To prevent static discharge that could ignite vapors

Q392. Which principle does a deluge sprinkler system operate on?

A. Open sprinklers activated by a detection system

B. Closed heads releasing individually

C. Constant low-pressure water misting

D. Sprinklers triggered by manual valve only

Q393. What type of fire extinguisher uses a sodium bicarbonate agent?

A. Dry chemical extinguisher

B. Wet chemical extinguisher

C. Halon extinguisher

D. Water extinguisher

Q394. Which standard provides guidance on combustible dust hazard management?

A. NFPA 652

B. NFPA 10

C. NFPA 13

D. NFPA 54

Q395. Which method best prevents flash fires in flammable liquid handling areas?

A. Using larger storage tanks

B. Adequate ventilation and vapor control

C. Reducing humidity

D. Minimizing signage

Q396. What determines the spacing between fire hydrants around a facility?

A. Fire flow demand and accessibility

B. Landscape design

C. Property ownership

D. Parking layout

Q397. Which factor most influences fire spread in high-rise buildings?

A. Stack effect

B. Ceiling height

C. Number of elevators

D. Building age

Q398. Which feature differentiates a clean agent system from CO_2 systems?

A. Lower installation cost

B. Higher extinguishing temperature

C. Safe for occupied spaces

D. Requires manual operation

Q399. Which OSHA document provides fire service feature guidance for building design?

A. OSHA 3256

B. OSHA 3071

C. OSHA 1910.147

D. OSHA 1910.1200

Q400. What is the principal advantage of an early suppression fast response (ESFR) sprinkler system?

A. Rapid discharge and high water flow for early fire knockdown

B. Reduced pipe size and water pressure

C. Eliminates need for detection systems

D. Operates only at high temperatures

Answer keys and Explanations (301-400)

Q301. Answer: B

Explanation: Adjustable-height platforms reposition the worker to reduce the need for overhead reaching, effectively lowering arm elevation and shoulder load. Engineering controls like height adjustment reduce static muscular effort, minimizing fatigue and risk of rotator cuff disorders. This solution ensures that the work zone remains within the optimal shoulder-to-elbow range for ergonomic safety.

Q302. Answer: A

Explanation: A neutral wrist posture keeps the hand, wrist, and forearm in a straight line without deviation or flexion. This alignment minimizes compressive and shear forces on tendons and nerves, reducing the likelihood of repetitive strain injuries like carpal tunnel syndrome. Proper wrist posture is a cornerstone of ergonomic design in typing and tool use.

Q303. Answer: B

Explanation: The ACGIH and ISO guidelines recommend a daily exposure limit of approximately 2.5 m/s^2 for hand–arm vibration. This limit helps prevent Hand-Arm Vibration Syndrome (HAVS), which can cause numbness, vascular constriction, and nerve damage. Maintaining vibration levels below this threshold protects workers operating powered hand tools and vibrating machinery.

Q304. Answer: C

Explanation: Isometric contractions involve sustained muscle tension without movement, restricting blood flow and oxygen delivery to muscle fibers. This leads to rapid fatigue and discomfort, especially in the shoulders and back. Ergonomic interventions aim to replace static postures with dynamic movements to improve circulation and reduce muscular strain during work tasks.

Q305. Answer: C

Explanation: The ergonomic reach envelope is the area a worker can comfortably reach without leaning or twisting. It defines the normal working area for task and control placement. Designing within this space ensures efficiency and minimizes awkward postures, promoting natural arm movement and reducing risk of strain during repetitive or precision work.

Q306. Answer: B

Explanation: Designing for extremes and adjustability ensures inclusivity across a broad range of users. For example, minimum reach is based on the 5th percentile female, and maximum clearance on the 95th percentile male. This approach guarantees comfort and safety for nearly all users, reducing the risk of exclusion and ergonomic mismatch.

Q307. Answer: B

Explanation: CTDs develop when repetitive and forceful motions continuously stress muscles, tendons, and nerves without sufficient recovery time. The combination of repetition and force accelerates fatigue and inflammation. Job redesign that reduces force requirements or integrates automation effectively lowers the risk of developing chronic musculoskeletal injuries.

Q308. Answer: D

Explanation: Angled microscope stands or tilting eyepieces allow users to maintain an upright head and neck posture, minimizing cervical flexion. Engineering solutions like these reduce static muscle load in the neck and shoulders. Adjustable visual equipment aligns the field of vision with natural posture, improving comfort and precision during detailed visual tasks.

Q309. Answer: B

Explanation: A pistol-grip screwdriver aligns the handle with the forearm when working at chest height, maintaining neutral wrist posture. This reduces ulnar and radial deviation, a common cause of wrist fatigue and tendon disorders. Tool selection based on work height and direction ensures ergonomic compatibility between hand position and applied force.

Q310. Answer: B

Explanation: The NIOSH Lifting Equation incorporates coupling quality—how well a worker can grasp and control the load—as a critical determinant of lifting frequency limits. Good coupling reduces required grip force and improves control, lowering biomechanical stress on the back and upper limbs. Poor coupling significantly increases injury risk in manual handling tasks.

Q311. Answer: C

Explanation: A slight recline of 100–110° reduces spinal compression and supports the lumbar curve. This posture distributes upper body weight more evenly across the spine, alleviating pressure on the intervertebral discs. Ergonomic chairs incorporate adjustable backrests and lumbar support to maintain neutral spine curvature for prolonged seated tasks.

Q312. Answer: B

Explanation: Anthropometric data guide control and display placement by defining reach and sight ranges for users. Panels designed based on reach envelopes and line-of-sight angles ensure comfortable operation and quick response. Properly located controls enhance usability and reduce operator fatigue, optimizing human–machine interaction efficiency and safety.

Q313. Answer: C

Explanation: Fine precision work benefits from a slightly elevated working height—approximately 5–10 cm above elbow level—to support visual focus and hand control. This minimizes the need for excessive neck flexion or arm elevation. Adjustable benches or platforms allow customization for precision assembly or inspection tasks, enhancing ergonomic posture.

Q314. Answer: B

Explanation: Interface complexity increases the likelihood of human error by overloading cognitive processing. Simplified layouts, consistent labeling, and intuitive feedback reduce confusion and decision time. Designing user-centered interfaces improves operational safety and performance, supporting principles of human factors engineering in high-reliability systems.

Q315. Answer: B

Explanation: Lifting objects above shoulder height compresses the rotator cuff tendons beneath the acromion, leading to shoulder impingement. Repetitive overhead motion increases tendon wear and inflammation. Ergonomic interventions focus on lowering storage shelves and using lifting aids to keep loads within shoulder height, preventing impingement syndromes.

Q316. Answer: A

Explanation: Color redundancy and logical grouping organize information for rapid visual processing. Consistent color coding improves recognition and reduces misinterpretation, especially in safety-critical systems. This principle follows human factors design standards, ensuring operators can detect, interpret, and respond efficiently to displayed data under time pressure.

Q317. Answer: B

Explanation: Anti-fatigue mats promote subtle leg muscle movement that enhances blood flow, reducing static load and discomfort during prolonged standing. They cushion impact and decrease lower limb strain. Combined with footrests and adjustable work surfaces, these mats significantly improve comfort and lower musculoskeletal stress in industrial and retail settings.

Q318. Answer: B

Explanation: Split keyboards allow wrists and forearms to align naturally, reducing ulnar deviation and tendon loading. This ergonomic design decreases strain on the carpal tunnel region, promoting comfort during extended typing. Coupled with proper chair height and hand positioning, split keyboards significantly lower repetitive strain risk in office environments.

Q319. Answer: D

Explanation: Diffuse ambient lighting minimizes glare and shadow formation, maintaining balanced luminance between screen and background. This reduces eye fatigue and squinting. Ergonomic lighting design combines indirect illumination and adjustable task lighting to support visual comfort, concentration, and productivity in computer-intensive work environments.

Q320. Answer: B

Explanation: The frequency multiplier (FM) in the NIOSH Lifting Equation adjusts the Recommended Weight Limit (RWL) based on how often lifts occur. Higher lifting frequencies reduce allowable load to prevent fatigue and cumulative trauma. This ensures that repetitive manual handling stays within safe physiological and biomechanical limits for workers.

Q321. Answer: B

Explanation: Cognitive ergonomics studies how humans perceive, remember, and respond to information. It aims to optimize system interfaces, workload, and feedback to support decision-making accuracy. Applications include control room design and software usability, ensuring that human cognitive capabilities align with task demands for reliable performance.

Q322. Answer: D

Explanation: Lack of stimulation reduces arousal and attentiveness, leading to vigilance decrement. Prolonged monotonous monitoring lowers cognitive engagement and reaction speed. Ergonomic solutions include task rotation, periodic feedback, and automated alerts that sustain operator alertness and reduce human error probability in continuous observation roles.

Q323. Answer: B

Explanation: Distinct and contextually relevant sounds help operators quickly recognize the source and meaning of alarms. Differentiated tone patterns and repetition rates improve detection and response accuracy. This principle ensures auditory signals are intuitive and not confused with ambient noise, enhancing situational awareness and safety response.

Q324. Answer: C

Explanation: Seated eye height measures the vertical distance from the seat reference point to the eyes when sitting upright. This dimension determines proper monitor height, display placement, and visual field design. Using accurate anthropometric data prevents excessive neck flexion and ensures comfortable viewing in seated work positions.

Q325. Answer: B

Explanation: A negative keyboard tilt angles the keyboard away from the user, allowing the wrists and forearms to remain in a neutral position rather than pronated. This reduces strain on the forearm muscles and tendons. Combined with proper desk height, this adjustment enhances comfort and prevents repetitive strain injuries in computer-based tasks.

Q326. Answer: C

Explanation: Grip span determines how well a worker can apply force without excessive muscle effort or awkward wrist angles. For repetitive work, maintaining an optimal grip span (typically 1.25–2 inches) allows the hand to generate sufficient force while minimizing strain. Proper handle sizing distributes pressure evenly across the palm, reducing fatigue, improving comfort, and preventing long-term musculoskeletal stress on the forearm and wrist.

Q327. Answer: B

Explanation: Minimizing horizontal reach distance significantly reduces moment arm length and spinal loading. When the load is kept close to the body, the mechanical advantage improves, and the required lifting force decreases. This principle is essential in NIOSH and OSHA ergonomic guidelines to prevent lower back strain and ensure safer manual material handling through proper load placement and workstation arrangement.

Q328. Answer: B

Explanation: An ergonomic job hazard analysis identifies risk factors such as repetition, force, awkward posture, and vibration within a job task. By systematically observing and documenting work activities, ergonomists can determine where redesign or control measures are needed. This proactive assessment helps prevent musculoskeletal disorders and promotes safer, more efficient work practices aligned with ergonomic and safety engineering principles.

Q329. Answer: A

Explanation: Elbow height provides the most practical reference for determining appropriate workbench height. Tasks requiring manual precision are typically set slightly above elbow level, while heavy or forceful work is placed below. Designing based on elbow height ensures neutral arm posture and minimizes shoulder elevation, reducing strain and promoting efficiency in both seated and standing workstation setups.

Q330. Answer: C

Explanation: Designing for user adjustability accommodates Variability in human body dimensions and preferences. Adjustable seats allow fine-tuning of height, backrest angle, and lumbar support, ensuring neutral posture and comfort. This flexibility enhances user satisfaction, minimizes static loading, and prevents lower back pain, reinforcing the ergonomic principle of fitting the workplace to the individual rather than the reverse.

Q331. Answer: A

Explanation: Arranging tools and materials based on frequency of use reduces unnecessary reaching and motion, improving efficiency and safety. Frequently used items should be within the primary reach zone, while less-used tools are placed farther away. This principle aligns with human factors engineering goals—minimizing motion waste and fatigue while enhancing workflow and productivity in repetitive or precision tasks.

Q332. Answer: B

Explanation: Sit-stand workstations promote posture variation by allowing workers to alternate between sitting and standing. This reduces static muscle load, improves circulation, and decreases the risk of musculoskeletal discomfort associated with prolonged sitting or standing. Encouraging movement throughout the day maintains flexibility, supports spinal health, and enhances comfort and focus during extended work periods.

Q333. Answer: B

Explanation: Maintaining neutral wrist alignment reduces compressive forces on nerves and tendons, preventing repetitive strain injuries such as carpal tunnel syndrome. Neutral wrist posture aligns the hand with the forearm, optimizing force transfer and comfort. Ergonomic tools and adjustable handle angles help maintain this alignment, promoting long-term musculoskeletal health and operational efficiency during repetitive manual tasks.

Q334. Answer: A

Explanation: High contrast and minimal clutter improve readability and reduce cognitive load in visual displays. Clear, well-organized information supports quick recognition and decision-making, essential in control environments. Human factors design standards emphasize visibility, consistency, and intuitive grouping, ensuring operators can rapidly detect anomalies and maintain high accuracy in monitoring and response tasks.

Q335. Answer: A

Explanation: The vertical distance of lift directly influences the mechanical load on the lumbar spine. The farther a worker bends, the greater the lever arm and resulting compressive force on spinal discs. Reducing the lift distance—by raising the origin height or lowering the destination—minimizes lower back stress and aligns with ergonomic lifting guidelines to prevent musculoskeletal disorders.

Q336. Answer: B

Explanation: Maintaining proper lighting contrast and monitor positioning minimizes eye strain by balancing luminance levels between the screen and surroundings. The monitor should be positioned 50–70 cm away, slightly below eye level, and free from glare sources. These adjustments support ocular comfort, reduce visual fatigue, and maintain productivity in extended computer-based work sessions.

Q337. Answer: B

Explanation: RULA (Rapid Upper Limb Assessment) and REBA (Rapid Entire Body Assessment) are validated tools for quantifying ergonomic risk associated with posture, movement, and load. They identify high-risk positions and guide targeted interventions. These methods support proactive ergonomics by providing structured, repeatable analysis for optimizing job design and reducing musculoskeletal disorder risk.

Q338. Answer: D

Explanation: Frictional handle materials, such as textured rubber, improve grip stability and control accuracy. This reduces slippage, allowing the user to apply consistent force with less muscular effort. Ergonomic handle surfaces are designed to enhance tactile feedback, supporting precision, reducing fatigue, and maintaining comfort during repetitive or high-precision tool operations.

Q339. Answer: B

Explanation: Ergonomic training educates employees on proper posture, lifting techniques, and body mechanics to recognize and prevent risk factors. While not a substitute for engineering controls, training fosters awareness and encourages safe behavior. Combined with ergonomic design, it empowers workers to maintain musculoskeletal health and contributes to an overall culture of injury prevention and well-being.

Q340. Answer: B

Explanation: Balanced tools distribute weight evenly, preventing excessive wrist torque and forearm strain during repetitive motion. Proper weight distribution reduces the muscular effort needed for stabilization, improving comfort and control. Ergonomically balanced tools enhance productivity, lower fatigue, and prevent cumulative strain injuries, especially in long-duration or precision-handling operations.

Q341. Answer: A

Explanation: Adjustability and modularity ensure that control panels can accommodate users of different sizes, visual ranges, and reach distances. Adjustable displays and control surfaces enable safe, comfortable operation without overextension or awkward postures. This design flexibility aligns with ergonomic principles of inclusivity and promotes optimal performance for multiple operators in shared environments.

Q342. Answer: A

Explanation: Leverage inefficiency occurs when joints move away from their neutral alignment, reducing the mechanical advantage of muscles and increasing required effort. Deviations from neutral posture alter joint angles and leverage, leading to faster fatigue and greater injury risk. Maintaining neutral alignment maximizes efficiency and minimizes unnecessary muscular strain in repetitive and static tasks.

Q343. Answer: B

Explanation: Adjustable footrests allow workers to shift weight periodically between legs, promoting blood circulation and reducing static load in the lower limbs. This dynamic support helps prevent muscle fatigue and discomfort during prolonged standing. Combined with adjustable surface height and supportive flooring, it enhances ergonomic safety and worker endurance in industrial and retail environments.

Q344. Answer: B

Explanation: Raising the monitor so the top is at or slightly below eye level encourages upright neck posture and reduces forward head displacement. Proper screen height aligns the visual axis naturally, minimizing strain on the cervical spine. This adjustment, combined with correct chair and desk height, ensures balanced posture and prevents neck and shoulder discomfort.

Q345. Answer: C

Explanation: Stress and cognitive load directly impact performance under time constraints by reducing working memory and decision accuracy. Ergonomic system design should minimize unnecessary information processing and optimize interface simplicity. Supporting cognitive ergonomics under pressure ensures consistent operator performance and lowers error rates in high-demand or safety-critical environments.

Q346. Answer: B

Explanation: Anti-vibration mounts and dampers reduce transmission of mechanical energy from the tool to the hand and arm. Engineering controls like isolation grips and suspended tool systems effectively lower daily vibration exposure levels. These modifications prevent Hand-Arm Vibration Syndrome (HAVS) and promote long-term occupational health in workers using powered hand tools.

Q347. Answer: B

Explanation: Proper lumbar support maintains the natural curvature of the lower spine, preventing slouching and disc compression. Supporting the lumbar region reduces muscle tension and fatigue, promoting neutral alignment during extended sitting. Ergonomically designed chairs with adjustable lumbar depth help prevent chronic back pain and improve seated posture comfort.

Q348. Answer: D

Explanation: Biomechanical efficiency refers to using body mechanics that maximize force production while minimizing energy expenditure and stress on joints. Maintaining neutral posture and proper load alignment optimizes leverage, reduces muscle fatigue, and enhances task performance. Ergonomically efficient movements preserve worker endurance and prevent injury across repetitive or physically demanding tasks.

Q349. Answer: B

Explanation: Adjustable task lighting focuses illumination where needed, enhancing visual clarity and precision during detailed assembly. Proper lighting reduces eye strain, supports accurate hand-eye coordination, and improves quality control. Combining adjustable luminance and glare control aligns with ergonomic lighting standards for precision-based manufacturing or inspection environments.

Q350. Answer: A

Explanation: Designing user-centered systems ensures that process controls align with human perception, reaction time, and cognitive limits. Incorporating human factors into safety design minimizes operational error, improves situational awareness, and enhances decision-making reliability. This integration fosters safer, more intuitive systems where human operators and machines interact efficiently within defined ergonomic parameters.

Q351. Answer: D

Explanation: The chain reaction sustains combustion by continuously releasing energy from the oxidation process. Once ignition occurs, free radicals within the flame react with oxygen and fuel vapors to propagate burning. Removing any element of the tetrahedron—such as cooling, smothering, or fuel isolation—interrupts this chain reaction, effectively extinguishing the fire.

Q352. Answer: C

Explanation: Class C fires involve energized electrical sources where nonconductive extinguishing agents like carbon dioxide or dry chemicals are used to avoid shock hazards. Once de-energized, the fire is treated as its base material type. Proper identification of Class C hazards is vital to prevent electrical injury during suppression.

Q353. Answer: A

Explanation: NFPA 10 establishes minimum requirements for selecting, inspecting, maintaining, and recharging portable extinguishers. It ensures readiness and effectiveness for first-response firefighting. Regular inspections verify accessibility, pressure, and serviceability to provide rapid protection against small incipient fires.

Q354. Answer: A

Explanation: Under NFPA 30, flammable liquids possess a flash point below 100°F (37.8°C). This characteristic indicates that vapors can ignite readily at normal ambient temperatures. Identifying such liquids allows proper storage, ventilation, and control of ignition sources to prevent vapor ignition and explosions.

Q355. Answer: B

Explanation: Dry pipe systems are charged with pressurized air or nitrogen that holds a valve closed to prevent water from entering the pipes until activation. This design is ideal for environments exposed to freezing conditions, ensuring that water does not freeze in the piping network.

Q356. Answer: B

Explanation: Carbon dioxide extinguishers suppress fires primarily by displacing oxygen around the flame, effectively suffocating the combustion process. They also offer limited cooling through gas expansion. CO_2 is especially suitable for electrical or flammable liquid fires due to its non-conductive nature and residue-free discharge.

Q357. Answer: A

Explanation: Water-based extinguishers are universally color-coded red under NFPA and ANSI identification systems. This standardization ensures quick recognition during emergencies. Water extinguishers are designed primarily for Class A fires involving ordinary combustibles such as wood, paper, and textiles.

Q358. Answer: A

Explanation: Class D fires involve combustible metals such as magnesium, titanium, sodium, and potassium. These fires require special extinguishing agents, often dry powders, that absorb heat and smother without reacting violently with the burning metal. Using water or foam can intensify such fires.

Q359. Answer: A

Explanation: Combustible dust presents an explosion hazard when fine particles become airborne and ignite within a confined space. Adequate ventilation, housekeeping, and ignition control prevent accumulation. NFPA 652 outlines management of combustible dust to minimize catastrophic explosion potential.

Q360. Answer: A

Explanation: Photoelectric smoke detectors operate by sensing light scatter or blockage caused by smoke particles entering the sensing chamber. They are highly responsive to smoldering fires that produce visible smoke before flames develop. Early warning allows timely evacuation and suppression actions.

Q361. Answer: C

Explanation: NFPA 13 specifies the design, installation, and testing standards for automatic sprinkler systems. It defines water supply, coverage, and system types to ensure reliable fire suppression performance. Compliance ensures optimal protection for life and property.

Q362. Answer: A

Explanation: Wet chemical extinguishers contain potassium acetate that cools and creates a soapy layer over burning oils, preventing re-ignition. This makes them ideal for Class K kitchen fires involving cooking fats and oils. They provide effective saponification and surface sealing.

Q363. Answer: A

Explanation: Fire walls are structural barriers with specific fire-resistance ratings designed to confine a fire to a defined area, preventing horizontal or vertical spread. These passive fire protection features enhance containment and safeguard adjacent sections of the structure.

Q364. Answer: B

Explanation: Identifying potential fire hazards is the foundation of a fire risk assessment. It involves systematically reviewing materials, processes, and conditions that could lead to ignition or fire spread. This allows proactive mitigation through controls, training, and system improvements.

Q365. Answer: A

Explanation: Static electricity control prevents sparks that can ignite flammable vapors, especially in handling or transferring flammable liquids. Grounding and bonding ensure charge dissipation, maintaining safe environments per NFPA 77 guidelines.

Q366. Answer: A

Explanation: OSHA 29 CFR 1910 Subpart L sets forth the requirements for fire protection systems, extinguishers, and emergency preparedness in general industry workplaces. Compliance ensures adequate detection, suppression, and evacuation capabilities to protect workers and property.

Q367. Answer: C

Explanation: UV/IR flame detectors identify specific wavelengths of radiation emitted by flames, enabling rapid detection in environments with open processes or flammable gases. These detectors enhance early warning where smoke-based devices are ineffective.

Q368. Answer: A

Explanation: Clean agent systems discharge gases like FM-200 or Inergen to displace oxygen or interrupt the chemical chain reaction, extinguishing fire without residue. They are used where water damage is unacceptable, such as in data centers or museums.

Q369. Answer: A

Explanation: Self-closing metal containers isolate oily rags from oxygen and heat accumulation, preventing spontaneous ignition from exothermic oxidation. Proper disposal and storage according to NFPA 30 and OSHA standards eliminate this common workplace hazard.

Q370. Answer: A

Explanation: Wet pipe sprinklers maintain pressurized water directly in the piping, releasing instantly when the heat-sensitive element (bulb or fusible link) activates. This immediate response provides reliable protection in areas without freezing risk.

Q371. Answer: D

Explanation: Yellow is the standard color for flammable liquid storage cabinets under NFPA 30 and OSHA guidelines. The bright color enhances visibility and hazard recognition, signaling flammable material storage to employees and emergency responders.

Q372. Answer: A

Explanation: Aqueous Film Forming Foam (AFFF) forms a thin film that spreads across hydrocarbon fuels, separating oxygen and suppressing vapor release. Its quick knockdown capability makes it ideal for petroleum-based fires in storage or processing areas.

Q373. Answer: A

Explanation: Air sampling smoke detectors continuously draw air through pipes and analyze particulates, detecting smoke at the earliest possible stage. This provides critical early warning in sensitive or high-value areas like data centers and laboratories.

Q374. Answer: C

Explanation: Adequate separation distances reduce radiant heat transfer between tanks, minimizing the risk of fire spread and explosion escalation. NFPA 30 defines spacing to protect adjacent units and structures, ensuring containment of potential fires.

Q375. Answer: A

Explanation: FM-200 (HFC-227ea) is a clean, non-conductive, and residue-free extinguishing agent that rapidly suppresses fires by absorbing heat and interrupting chemical reactions. It's safe for occupied areas and widely used in electronic equipment protection.

Q376. Answer: A

Explanation: A preaction sprinkler system prevents accidental water discharge in sensitive environments like museums, data centers, or archives. It uses both detection and activation signals before water enters the pipes, combining dry pipe features with added verification. This dual control ensures water is released only when a fire is confirmed, protecting valuable assets from unnecessary water damage while maintaining reliable fire suppression capabilities under controlled conditions.

Q377. Answer: A

Explanation: NFPA 72 outlines the design, installation, performance, and maintenance criteria for fire alarm systems. It provides detailed requirements for detection devices, signaling, annunciation, and system integrity. Adherence ensures prompt notification during emergencies, allowing effective evacuation and response. This standard supports interoperability and enhances life safety by maintaining reliability across integrated fire detection and alarm networks used in all types of facilities.

Q378. Answer: B

Explanation: Eliminating ignition sources such as sparks, static electricity, or open flames provides the most effective preventive control in flammable gas storage areas. This includes grounding, bonding, and using intrinsically safe electrical equipment. By removing potential ignition energy, facilities reduce the likelihood of explosions and maintain compliance with NFPA 55 and OSHA standards that prioritize source control over reliance on suppression or reactive measures.

Q379. Answer: A

Explanation: Explosion-proof exhaust systems safely remove flammable vapors from storage areas, maintaining concentrations below their lower explosive limit. These systems use spark-resistant motors and ducting to prevent ignition within the airflow path. Proper ventilation reduces vapor accumulation, ensuring compliance with NFPA 30 and OSHA 1910 standards for safe flammable liquid storage and minimizing the risk of vapor ignition and subsequent fire events.

Q380. Answer: A

Explanation: Class K extinguishers are specifically designed to combat fires involving cooking oils and fats found in commercial kitchens. They release a fine mist of potassium acetate that reacts with burning oils to create a soap-like barrier, cooling the surface and preventing re-ignition. This process, known as saponification, effectively seals vapors and minimizes oxygen contact, ensuring complete and safe fire suppression without flashback risks.

Q381. Answer: C

Explanation: Foam-based suppression systems form a continuous film or blanket over the fuel surface, effectively separating it from oxygen and blocking vapor release. This action starves the fire and prevents re-ignition. Commonly used in petrochemical facilities and aircraft hangars, these systems comply with NFPA 11, offering rapid coverage and control of Class B fires involving flammable and combustible liquids.

Q382. Answer: A

Explanation: Flash point is the lowest temperature at which a liquid produces sufficient vapor to form an ignitable mixture in air. It is a critical indicator of fire risk used to classify flammable and combustible liquids under NFPA 30. Lower flash points indicate greater hazard potential, guiding safe handling, storage, and ventilation practices in industrial and laboratory settings where fire prevention is essential.

Q383. Answer: D

Explanation: Fire drills evaluate employee response times and coordination during emergency evacuations. They help identify procedural gaps, improve route familiarity, and strengthen confidence under pressure. Regular drills also ensure that alarm systems, communication protocols, and assembly point procedures function effectively, aligning with OSHA 1910.38 and NFPA 101 requirements for effective emergency preparedness and employee safety culture enhancement.

Q384. Answer: A

Explanation: Carbon dioxide extinguishers effectively suppress Class B fires involving flammable liquids by displacing oxygen and cooling the flame zone. Their non-conductive nature also makes them safe for electrical equipment. CO_2 leaves no residue, preventing contamination and ensuring quick restoration of operations, aligning with NFPA 10 guidelines for portable extinguisher selection and use across industrial and laboratory environments.

Q385. Answer: B

Explanation: Intumescent coatings swell when exposed to high temperatures, forming a thick char layer that insulates structural steel from heat exposure. This thermal barrier delays temperature rise and maintains structural integrity during fires, providing additional evacuation time. These coatings are key components in passive fire protection systems as defined in NFPA 251 and modern building codes emphasizing fire endurance.

Q386. Answer: A

Explanation: FM-200 (HFC-227ea) is widely used in clean agent fire suppression for data centers and critical facilities. It extinguishes flames rapidly through heat absorption and chemical interference, leaving no residue and posing minimal risk to electronics or personnel. FM-200 complies with NFPA 2001, providing environmentally safer alternatives to Halon-based systems while maintaining high suppression efficiency.

Q387. Answer: A

Explanation: Combustible dust hazards are mitigated through containment systems equipped with explosion vents or suppression devices that relieve pressure and direct energy safely. This design minimizes structural damage and prevents chain-reaction explosions. Following NFPA 68 and NFPA 652 ensures proper dust management, venting design, and system isolation to protect personnel and equipment in industrial environments.

Q388. Answer: C

Explanation: Fire load quantifies the total heat energy that could be released during a fire, based on combustible contents and structural materials. It helps determine suppression requirements and structural fire resistance. Measured in MJ/m^2, it guides design for sprinkler density, ventilation, and passive fire protection in compliance with NFPA and building code standards.

Q389. Answer: A

Explanation: Fire dampers close automatically when a duct reaches a specific temperature, preventing fire and smoke from traveling between building compartments. They are integral to HVAC fire safety, required under NFPA 90A and building codes. Regular inspection ensures functionality and containment integrity during emergencies, protecting escape routes and adjoining areas from smoke propagation.

Q390. Answer: A

Explanation: Water supply systems for fire suppression depend on available pressure and flow rate to maintain consistent performance across sprinkler networks. NFPA 13 and NFPA 24 require hydraulic calculations to verify sufficient water delivery under peak demand. Reliable pressure ensures sprinkler coverage, system activation, and continuous suppression capability during extended fire conditions.

Q391. Answer: D

Explanation: Grounding and bonding eliminate static electricity accumulation by connecting conductive objects to a common ground. This prevents sparks that could ignite flammable vapors during liquid transfer or mixing. Adherence to NFPA 77 ensures effective electrostatic control, reducing fire and explosion risks in industrial operations involving volatile substances or combustible gases.

Q392. Answer: A

Explanation: Deluge systems use open sprinklers connected to a dry network that fills with water once a separate detection system activates. This delivers rapid, full-area discharge suitable for high-hazard areas such as chemical processing plants or aircraft hangars. NFPA 15 governs their design to ensure quick knockdown capability in large-scale flammable liquid fires.

Q393. Answer: A

Explanation: Dry chemical extinguishers containing sodium bicarbonate interrupt the chemical reaction of Class B and C fires. The agent releases carbon dioxide upon heating, smothering the flame and preventing re-ignition. These extinguishers provide fast, effective coverage and are commonly used in industrial and vehicular environments as recommended by NFPA 10.

Q394. Answer: A

Explanation: NFPA 652 serves as the foundational standard for managing combustible dust hazards across industries. It covers dust hazard analysis (DHA), control measures, and explosion protection requirements. Compliance helps facilities prevent catastrophic events through effective housekeeping, ignition control, and dust collection system maintenance.

Q395. Answer: B

Explanation: Proper ventilation and vapor control prevent the buildup of flammable atmospheres by maintaining concentrations below lower explosive limits. Local exhaust, vapor recovery systems, and well-sealed containers reduce ignition risk. NFPA 30 mandates these controls to ensure safe operations and limit flash fire hazards in liquid transfer and storage processes.

Q396. Answer: A

Explanation: Fire hydrant spacing depends on fire flow requirements, building occupancy risk, and response accessibility. NFPA 1 and local codes specify placement to ensure adequate water supply for suppression activities. Proper spacing facilitates quick fire department access, maintaining sufficient pressure for hose lines and firefighting efficiency across all hazard zones.

Q397. Answer: A

Explanation: Stack effect refers to vertical air movement within tall structures due to temperature differences between inside and outside air. During a fire, it can cause rapid smoke and heat movement through shafts or stairwells. Proper compartmentation, pressurization, and smoke control systems mitigate this effect, protecting evacuation routes and improving fire containment.

Q398. Answer: C

Explanation: Clean agent systems like FM-200 and Novec 1230 are designed for occupied spaces because they extinguish fires without reducing oxygen to unsafe levels. CO_2 systems, while effective, can pose asphyxiation hazards. Clean agents leave no residue and preserve electronic equipment integrity, making them ideal for critical facilities like control rooms or server environments.

Q399. Answer: A

Explanation: OSHA 3256, "Fire Service Features of Buildings and Fire Protection Systems," offers design recommendations for fire department access, water supply, and system integration. It ensures facilities accommodate safe emergency response operations. Following these guidelines improves coordination, reduces hazards to responders, and enhances structural safety under fire conditions.

Q400. Answer: A

Explanation: ESFR sprinklers deliver high water discharge density with large droplets, designed to suppress fires at their source before significant growth. Commonly used in warehouses, these systems protect high-piled storage without in-rack sprinklers. NFPA 13 recognizes ESFR systems for their quick activation and superior fire control performance in demanding storage applications.

Questions (401-500)

Q401. What is the main purpose of a fire alarm control panel (FACP)?

A. To provide manual suppression

B. To operate sprinkler discharge valves

C. To regulate building temperature

D. To monitor and manage fire detection and alarm signals

Q402. Which type of detection system identifies heat by measuring the rate of temperature increase?

A. Rate-of-rise heat detector

B. Fixed temperature detector

C. Photoelectric detector

D. Ionization detector

Q403. Which hazard is most associated with hot work operations?

A. Ignition of nearby combustible materials

B. Compressed gas expansion

C. Electrical overload

D. Noise exposure

Q404. What is the correct extinguishing agent for a magnesium fire?

A. CO_2 extinguisher

B. Dry powder extinguisher

C. Water extinguisher

D. Foam extinguisher

Q405. Which system ensures fire detection continuity during power outages?

A. Battery backup and secondary power supply

B. Manual activation

C. Water reservoir

D. Fire wall integration

Q406. Which parameter defines the lower explosive limit (LEL) of a gas?

A. Minimum concentration of gas in air capable of ignition

B. Highest temperature at which gas vaporizes

C. Lowest oxygen concentration needed for combustion

D. Temperature when flame ceases

Q407. What is the purpose of a flame arrester in a flammable liquid storage system?

A. To equalize vapor pressure

B. To detect flame presence

C. To prevent flame propagation back into the container

D. To monitor temperature

Q408. Which fire suppression agent works by interrupting the chemical chain reaction?

A. Halon

B. Water

C. Foam

D. Sand

Q409. What is the most reliable indicator of a smoldering fire in a storage area?

A. UV/IR sensor

B. Heat detector

C. Flame detector

D. Photoelectric smoke detector

Q410. Which NFPA standard addresses installation of standpipe and hose systems?

A. NFPA 14

B. NFPA 13

C. NFPA 25
D. NFPA 30
Q411. What does the fire resistance rating of a wall indicate?
A. Duration the wall resists fire exposure before failure
B. Thickness of insulation applied
C. Flame color during exposure
D. Temperature rise limit on surface
Q412. Which extinguishing agent is most appropriate for Class B fires?
A. Water
B. Foam
C. Dry powder
D. Sand
Q413. Which document outlines fire safety requirements for laboratories using flammable materials?
A. NFPA 45
B. NFPA 10
C. NFPA 58
D. NFPA 13
Q414. Which fire protection feature activates automatically to contain smoke at ceiling level?
A. Fire door
B. Fire damper
C. Smoke curtain
D. Flashing barrier
Q415. Which fire alarm component converts heat or smoke into an electrical signal?
A. Detector
B. Control panel
C. Strobe light
D. Manual pull station
Q416. Which maintenance activity ensures sprinkler reliability?
A. Regular inspection, testing, and maintenance (ITM)
B. Annual repainting
C. Flow reduction testing only
D. Visual check from distance
Q417. What is the purpose of a fire command center in high-rise buildings?
A. To control building lighting
B. To house fire extinguishers
C. To serve as a mechanical storage area
D. To coordinate all emergency communication and control systems
Q418. What distinguishes a wet chemical suppression system used in kitchens?
A. Its ability to cool and create a soap-like layer over burning oils
B. Its use of carbon dioxide discharge
C. Its constant dry powder release
D. Its use for metal fires
Q419. Which element must be eliminated to extinguish a fire according to the fire tetrahedron?
A. Only oxygen
B. Only heat
C. Any one of heat, fuel, oxygen, or chemical reaction
D. Both heat and oxygen
Q420. Which system is used to detect heat or smoke in hazardous classified areas?
A. Explosion-proof detection system
B. Standard open detector

C. Infrared camera
D. Manual alarm only
Q421. Which NFPA code addresses installation and testing of fire pumps?
A. NFPA 14
B. NFPA 13
C. NFPA 25
D. NFPA 20
Q422. What is the main advantage of a double interlock preaction system?
A. It requires both detection and sprinkler activation before discharge
B. It operates manually only
C. It continuously sprays water
D. It depends solely on temperature changes
Q423. Which term defines the automatic closing of doors upon alarm activation?
A. Deadbolt lock
B. Manual latch
C. Self-closing mechanism
D. Panic hardware
Q424. What is the role of thermal insulation in passive fire protection?
A. To delay heat transfer and structural failure
B. To increase airflow
C. To absorb smoke
D. To suppress flames directly
Q425. Which detection device uses ionized air to sense invisible combustion particles?
A. Ionization smoke detector
B. Photoelectric detector
C. Flame detector
D. Thermal sensor
Q426. What is the primary function of a deluge sprinkler system?
A. To provide localized water discharge
B. To deliver water simultaneously through all open nozzles
C. To detect fire using smoke sensors
D. To maintain air pressure within pipes
Q427. Why are Class D fire extinguishers specifically designed for metal fires?
A. They contain dry powder that smothers and isolates burning metal
B. They use carbon dioxide for rapid cooling
C. They release foam to coat surfaces
D. They rely on water spray for extinguishment
Q428. Which hazard is controlled by grounding and bonding of equipment?
A. Static electricity buildup
B. Excessive vibration
C. Corrosion
D. Noise emission
Q429. Which device provides an early warning of fire by sensing changes in infrared radiation?
A. Gas sensor
B. Smoke detector
C. Heat detector
D. Flame detector
Q430. What is the main purpose of fire compartmentation in building design?
A. To prevent fire and smoke from spreading between areas
B. To store firefighting equipment

C. To allow natural ventilation
D. To reduce building weight
Q431. Which type of sprinkler system contains both air and water separated by a valve?
A. Preaction system
B. Dry pipe system
C. Wet pipe system
D. Deluge system
Q432. What is the most important step in preventing combustible dust explosions?
A. Effective housekeeping and dust control
B. Increasing air humidity
C. Reducing lighting intensity
D. Using metal floors
Q433. What does a fire load calculation determine?
A. Total potential heat energy of combustibles in an area
B. Air temperature variation
C. Water supply flow rate
D. Oxygen level
Q434. Why are fire dampers installed in HVAC systems?
A. To reduce noise in airflows
B. To filter air from contaminants
C. To block fire and smoke from spreading through ducts
D. To regulate air pressure
Q435. Which method is most effective for preventing spontaneous combustion in oily rags?
A. Storing in covered metal containers
B. Leaving them in open piles
C. Washing with cold water only
D. Drying them under sunlight
Q436. What is the function of a fire-rated door assembly?
A. To provide soundproofing
B. To delay fire and smoke spread while maintaining egress
C. To regulate room pressure
D. To reduce vibration
Q437. Which feature distinguishes an automatic sprinkler from an open nozzle?
A. A heat-sensitive element sealing the water path
B. A larger spray pattern
C. A manual trigger mechanism
D. A constant airflow sensor
Q438. Why should exit routes be illuminated at all times?
A. To ensure visibility and safe evacuation during emergencies
B. To reduce power usage
C. To deter unauthorized entry
D. To meet aesthetic standards
Q439. What type of hazard is associated with improperly stored flammable liquids?
A. Mechanical stress
B. Increased humidity
C. Vapor ignition and explosion
D. Electrical interference
Q440. Which material is commonly used in firestopping applications?
A. Intumescent sealant
B. Rubber gasket

C. Silicone grease
D. Vinyl tape
Q441. What is the main advantage of a clean agent fire suppression system?
A. It extinguishes fires without leaving residue or damaging equipment
B. It uses high water pressure
C. It produces thick smoke
D. It relies on foam layering
Q442. Why is maintaining clear access to fire hydrants essential?
A. To allow public parking
B. To ensure firefighters can connect hoses quickly
C. To enhance landscaping
D. To prevent erosion
Q443. Which extinguisher type is suitable for energized electrical equipment fires?
A. Carbon dioxide extinguisher
B. Foam extinguisher
C. Water extinguisher
D. Wet chemical extinguisher
Q444. What is the main purpose of an emergency evacuation drill?
A. To practice safe and orderly evacuation procedures
B. To train on equipment maintenance
C. To check lighting brightness
D. To test building temperature
Q445. What is the role of an automatic fire alarm system?
A. To ventilate enclosed spaces
B. To suppress fires directly
C. To regulate building temperature
D. To detect fire conditions and alert occupants automatically
Q446. Which hazard can occur if sprinkler heads are painted or obstructed?
A. Delayed activation or failure to discharge
B. Improved sensitivity
C. Reduced corrosion
D. Increased water flow
Q447. Why should fire exits always open in the direction of travel?
A. To allow quick and safe egress during emergencies
B. To save floor space
C. To improve building ventilation
D. To align with aesthetic design
Q448. What is the main advantage of an addressable fire alarm system over a conventional system?
A. It operates without sensors
B. It uses simpler wiring
C. It identifies the exact location of the activated device
D. It triggers all alarms simultaneously
Q449. What is the main role of fire-resistant coatings on structural steel?
A. To delay heat absorption and maintain strength during fire
B. To improve appearance
C. To prevent corrosion only
D. To provide insulation from sound
Q450. What is the main objective of fire drills in workplaces?
A. To test the effectiveness of emergency response plans
B. To check power supply

C. To train employees for normal operations
D. To measure building height
Q451. Which U.S. law governs the cleanup of hazardous waste sites?
A. Clean Water Act
B. CERCLA
C. RCRA
D. EPCRA
Q452. Which agency primarily enforces air quality standards in the United States?
A. OSHA
B. NIOSH
C. EPA
D. DOT
Q453. What is the main purpose of a Spill Prevention, Control, and Countermeasure (SPCC) Plan?
A. To control air emissions
B. To prevent oil discharges into navigable waters
C. To reduce solid waste
D. To manage stormwater runoff
Q454. Which waste classification under RCRA refers to discarded commercial chemical products?
A. P-list and U-list wastes
B. F-list wastes
C. K-list wastes
D. D-list wastes
Q455. Which air pollutant is a key component of photochemical smog?
A. Sulfur dioxide
B. Carbon monoxide
C. Nitrogen oxides
D. Methane
Q456. Which control device is commonly used to remove particulate matter from industrial exhaust streams?
A. Cyclone separator
B. Catalytic converter
C. Absorption tower
D. Flare stack
Q457. What is the primary purpose of secondary containment for chemical storage tanks?
A. Improve visual inspection
B. Provide extra volume to capture spills or leaks
C. Reduce evaporation losses
D. Maintain tank temperature stability
Q458. Which process is used to reduce biological oxygen demand (BOD) in wastewater treatment?
A. Sedimentation
B. Filtration
C. Aeration
D. Neutralization
Q459. Which environmental program governs the management of hazardous waste from cradle to grave?
A. CERCLA
B. RCRA
C. EPCRA
D. TSCA
Q460. Which pollutant is most closely associated with acid rain formation?

A. Carbon dioxide
B. Methane
C. Sulfur dioxide
D. Ozone

Q461. Which component of an Environmental Management System (EMS) ensures continuous improvement?
A. Policy statement
B. Planning
C. Management review
D. Corrective action process

Q462. What is the main objective of environmental impact assessments (EIAs)?
A. To identify project funding
B. To evaluate potential environmental effects before project approval
C. To assess financial feasibility
D. To determine workforce needs

Q463. Which treatment method is typically used for volatile organic compound (VOC) removal from air streams?
A. Incineration
B. Filtration
C. Ion exchange
D. Sedimentation

Q464. Which federal law focuses on emergency planning and community right-to-know?
A. EPCRA
B. CERCLA
C. RCRA
D. TSCA

Q465. What is the function of a scrubber in an air pollution control system?
A. Capture particulate matter only
B. Remove gases and particulates using liquid contact
C. Burn pollutants at high temperature
D. Absorb noise emissions

Q466. Which process converts ammonia into nitrate during biological wastewater treatment?
A. Denitrification
B. Precipitation
C. Nitrification
D. Coagulation

Q467. What is the main goal of sustainability programs within organizations?
A. To maximize resource use
B. To reduce environmental footprint while maintaining productivity
C. To increase short-term profits
D. To minimize employee participation

Q468. Which emergency management phase includes recovery and restoration of normal operations?
A. Mitigation
B. Response
C. Recovery
D. Preparedness

Q469. Which monitoring method is used to detect airborne particulates in workplace environments?
A. Spectrophotometry
B. Gravimetric sampling
C. Gas chromatography

D. Ultraviolet analysis

Q470. Which regulation governs the management of toxic substances in commerce?

A. CERCLA

B. RCRA

C. TSCA

D. EPCRA

Q471. Which document outlines procedures for responding to hazardous waste incidents?

A. Spill Contingency Plan

B. Training Matrix

C. Hazard Analysis Summary

D. Environmental Audit Plan

Q472. Which practice helps minimize solid waste generation at the source?

A. Waste disposal

B. Pollution prevention

C. Incineration

D. Chemical treatment

Q473. Which atmospheric layer contains most weather-related phenomena?

A. Stratosphere

B. Troposphere

C. Mesosphere

D. Thermosphere

Q474. Which type of pollution results primarily from runoff containing fertilizers and pesticides?

A. Point source pollution

B. Non-point source pollution

C. Airborne particulate pollution

D. Industrial wastewater pollution

Q475. Which water quality indicator measures nutrient enrichment and potential for algal growth?

A. Dissolved oxygen

B. pH

C. Nitrate concentration

D. Temperature

Q476. Which federal law regulates the discharge of pollutants into U.S. waters?

A. CERCLA

B. RCRA

C. Clean Water Act

D. Safe Drinking Water Act

Q477. What is the main objective of the Clean Air Act?

A. Control hazardous waste generation

B. Regulate air emissions from stationary and mobile sources

C. Manage solid waste disposal

D. Establish workplace air quality standards

Q478. Which term describes water that infiltrates soil and collects contaminants before reaching groundwater?

A. Stormwater runoff

B. Leachate

C. Greywater

D. Effluent

Q479. Which component of an environmental audit ensures compliance verification?

A. Document review

B. Sampling and testing

C. Field observation
D. Corrective action review
Q480. Which process converts organic waste into a nutrient-rich soil amendment?
A. Composting
B. Incineration
C. Filtration
D. Chemical oxidation
Q481. Which environmental monitoring parameter indicates oxygen availability for aquatic life?
A. pH
B. Dissolved oxygen
C. Turbidity
D. Nitrate concentration
Q482. Which waste minimization technique focuses on reusing recovered materials in the production cycle?
A. Recycling
B. Incineration
C. Landfilling
D. Neutralization
Q483. Which environmental parameter is used to measure acidity or alkalinity in water?
A. Conductivity
B. Turbidity
C. pH
D. Nitrate
Q484. What is the function of a flocculation process in water treatment?
A. Kill microorganisms
B. Remove fine suspended solids
C. Neutralize acids
D. Reduce dissolved gases
Q485. Which term refers to the maximum concentration of a contaminant permitted in drinking water?
A. Action level
B. Threshold limit value
C. Maximum contaminant level
D. Exposure limit
Q486. Which step in waste management hierarchy has the highest environmental priority?
A. Treatment
B. Disposal
C. Source reduction
D. Recycling
Q487. Which pollutant is primarily responsible for ozone layer depletion?
A. Carbon dioxide
B. Chlorofluorocarbons
C. Sulfur dioxide
D. Methane
Q488. Which type of pollution occurs when heat from industrial discharges raises water temperature?
A. Chemical pollution
B. Thermal pollution
C. Radiological pollution
D. Organic pollution
Q489. What is the main advantage of using renewable energy sources?

A. They eliminate all pollution
B. They are cost-free to produce
C. They reduce dependence on finite fossil fuels
D. They require no maintenance
Q490. Which type of landfill liner provides the most effective leachate containment?
A. Clay only
B. Composite liner
C. Gravel layer
D. Sand filter
Q491. Which air pollution control device removes fine particles using electric charge?
A. Baghouse filter
B. Electrostatic precipitator
C. Cyclone separator
D. Wet scrubber
Q492. What is the purpose of the National Environmental Policy Act (NEPA)?
A. Mandate pollution cleanup programs
B. Require federal agencies to assess environmental effects before actions
C. Regulate industrial waste disposal
D. Monitor air and water standards
Q493. Which pollutant forms when sulfur dioxide and nitrogen oxides react with water vapor?
A. Ground-level ozone
B. Particulate matter
C. Carbon monoxide
D. Acid rain
Q494. Which environmental management practice helps reduce emissions from transportation fleets?
A. Increased idling
B. Use of low-emission or alternative fuel vehicles
C. Route expansion
D. Higher cargo loads
Q495. What is the primary purpose of environmental risk assessment?
A. Assign project costs
B. Identify potential environmental hazards and evaluate their impact
C. Measure public perception
D. Determine facility age
Q496. Which term refers to the intentional release of treated water into natural water bodies?
A. Discharge
B. Diffusion
C. Sedimentation
D. Containment
Q497. Which type of waste includes items like batteries and fluorescent bulbs?
A. Universal waste
B. Industrial waste
C. Municipal solid waste
D. Agricultural waste
Q498. Which factor primarily influences dispersion of air pollutants?
A. Air pressure
B. Wind speed and atmospheric stability
C. Solar radiation intensity
D. Humidity levels

Q499. Which waste management process involves burning waste at high temperatures to reduce volume?

A. Composting

B. Incineration

C. Recycling

D. Landfilling

Q500. What is the main benefit of implementing ISO 14001 Environmental Management Systems?

A. Reduce employee workload

B. Eliminate all environmental risks

C. Establish a structured approach for environmental improvement

D. Focus solely on waste reduction

Answer keys and Explanations (401-500)

Q401. Answer: D

Explanation:

The Fire Alarm Control Panel (FACP) functions as the main hub for managing signals from detectors and alarm devices. It processes input, triggers appropriate alerts, and coordinates with safety systems to ensure rapid emergency response. The panel's continuous monitoring and system integration provide essential protection, maintaining building readiness and ensuring that alarm activation leads to immediate and effective evacuation or suppression actions.

Q402. Answer: A

Explanation:

A rate-of-rise heat detector measures how quickly the temperature increases over time. When the temperature rises faster than a preset threshold, the detector activates an alarm, signaling the presence of fire. This method provides faster detection than fixed temperature sensors in rapidly developing fires, offering early warning and allowing quicker emergency responses to control potential hazards effectively.

Q403. Answer: A

Explanation:

Hot work operations such as welding, cutting, and grinding create sparks and heat that can easily ignite nearby combustible materials or vapors. The main hazard is uncontrolled fire spread from these ignition sources. Implementing a hot work permit system, maintaining fire watch personnel, and removing flammable materials from the area are crucial to prevent incidents during such tasks.

Q404. Answer: B

Explanation:

Combustible metal fires, including those involving magnesium, require a dry powder extinguisher. The powder absorbs heat and forms a crust over the burning metal, cutting off oxygen and stopping combustion. Using water or foam on metal fires can cause violent reactions and spread molten metal, increasing danger to personnel and property.

Q405. Answer: A

Explanation:

A reliable secondary power supply, such as batteries or generators, ensures fire detection systems remain operational during power failures. Continuous power is critical to maintain alarms, detection signals, and control functions. This redundancy provides uninterrupted protection, ensuring that emergency alerts and notifications function even when the primary power source is unavailable.

Q406. Answer: A

Explanation:

The Lower Explosive Limit (LEL) represents the minimum gas concentration in air that can ignite when exposed to a heat source. Below this limit, the gas mixture is too lean to burn. Maintaining gas concentrations below the LEL prevents explosions, particularly in storage or confined spaces where vapor buildup poses significant safety risks.

Q407. Answer: C

Explanation:

A flame arrester blocks flames from traveling back into containers or piping systems containing flammable vapors. Its internal mesh or metal element absorbs and disperses heat, cooling the flame front below ignition temperature. This device prevents explosions and ensures safe vapor management during storage and handling of volatile materials.

Q408. Answer: A

Explanation:

Halon extinguishes fires by interrupting the chemical chain reaction that sustains combustion. It quickly halts flame propagation, suppressing fire without residue. This agent was widely used for electronic and aviation applications before being replaced by environmentally safer clean agents that perform similar functions while preserving sensitive equipment and preventing secondary damage.

Q409. Answer: D

Explanation:

Photoelectric smoke detectors detect visible smoke particles produced by slow, smoldering fires. They use light scatter or blockage to sense smoke and trigger alarms early. Their high sensitivity makes them ideal for storage or office environments where fires may develop gradually, allowing faster intervention and reducing potential property loss.

Q410. Answer: A

Explanation:

This standard outlines design, installation, and maintenance criteria for standpipe and hose systems used for manual firefighting. It defines pressure requirements, connection types, and testing procedures. Such systems ensure firefighters and trained occupants have immediate access to an effective water supply when suppression systems are not sufficient on their own.

Q411. Answer: A

Explanation:

A fire resistance rating measures how long a wall or structural component can resist exposure to fire without losing its integrity. This rating helps control fire spread, allowing time for evacuation and reducing damage. Fire-rated walls form essential passive protection that complements active suppression systems for comprehensive safety.

Q412. Answer: B

Explanation:

Foam extinguishers are best suited for Class B fires involving flammable liquids. The foam forms a barrier between the fuel and oxygen, suppressing vapors and preventing re-ignition. It also cools the fuel surface, making it an efficient and safe method for controlling flammable liquid fires in industrial and storage environments.

Q413. Answer: A

Explanation:

This document provides fire protection guidance for laboratories that use flammable and combustible substances. It includes recommendations on storage, ventilation, and fire prevention measures specific to laboratory environments. Following these requirements helps reduce the risk of fires, protect personnel, and preserve research equipment and materials effectively.

Q414. Answer: C

Explanation:

Smoke curtains deploy automatically during a fire to contain smoke within a specific area and prevent its spread through the building. They help maintain visibility in escape routes and allow occupants more time to evacuate. These curtains also improve smoke control performance by keeping smoke from descending into lower occupied spaces.

Q415. Answer: A

Explanation:

A detector converts physical evidence of fire, such as heat, smoke, or flame, into an electrical signal that alerts the control panel. This signal then triggers alarms or suppression systems. Reliable detectors provide early warning, enabling timely evacuation and effective firefighting response to prevent injury and property loss.

Q416. Answer: A

Explanation:

Routine inspection, testing, and maintenance ensure that sprinklers perform correctly during a fire. Regular checks of valves, water flow, and pressure confirm the system's readiness. These preventive activities detect corrosion, blockages, or mechanical issues early, maintaining full reliability of fire suppression capabilities and protecting occupants and property continuously.

Q417. Answer: D

Explanation:

The fire command center serves as the central control hub during emergencies. It integrates alarms, communication, ventilation, and building systems to support responders. This setup allows efficient coordination, quick decision-making, and better situational awareness during fires, ensuring that life safety operations proceed smoothly and effectively under emergency conditions.

Q418. Answer: A

Explanation:

Wet chemical systems release a potassium-based solution that reacts with burning cooking oils to form a thick, soapy film. This process cools the surface, seals vapors, and prevents re-ignition. Designed for commercial kitchens, the system provides efficient suppression of grease fires while minimizing damage to surrounding equipment and ensuring food service safety.

Q419. Answer: C

Explanation:

The fire tetrahedron shows that removing any one of its four elements—heat, fuel, oxygen, or chemical chain reaction—will stop combustion. Fire control strategies are based on this concept, focusing on cooling, smothering, fuel isolation, or chemical interruption to extinguish fires safely and effectively in various situations.

Q420. Answer: A

Explanation:

Explosion-proof detection systems are specifically designed for environments where flammable gases or vapors may be present. These devices are sealed to prevent ignition of surrounding atmospheres and provide reliable detection without producing sparks. Their rugged construction ensures safe and continuous operation under extreme industrial conditions.

Q421. Answer: D

Explanation:

This code establishes standards for installing and maintaining fire pumps that supply pressure to suppression systems. It includes testing procedures to verify pump performance and reliability. Regular inspections and maintenance ensure pumps provide adequate water pressure for fire suppression operations during emergencies, protecting both life and property.

Q422. Answer: A

Explanation:

A double interlock preaction system activates only when both a detection signal and a sprinkler head trigger occur. This design minimizes the risk of accidental water discharge from false alarms or mechanical issues. It provides a safeguard for sensitive areas like computer rooms where water damage would cause significant operational disruption.

Q423. Answer: C

Explanation:

Self-closing mechanisms automatically shut fire-rated doors when an alarm activates, preserving compartmentation and slowing smoke and flame spread. These devices ensure escape routes remain protected during emergencies, providing critical time for evacuation and minimizing fire propagation between different building zones.

Q424. Answer: A

Explanation:

Thermal insulation slows heat transfer through structural elements, reducing the risk of collapse during prolonged fire exposure. It preserves the load-bearing capacity of materials, allowing safe evacuation and firefighting. Used with fire-resistant coatings and barriers, insulation forms part of a comprehensive passive protection strategy that enhances building resilience.

Q425. Answer: A

Explanation:

An ionization smoke detector uses a small radioactive source to ionize air within a chamber. When combustion particles enter, they disrupt the electrical current, triggering the alarm. This technology is highly effective for detecting fast-flaming fires with small invisible particles, providing early warning and critical response time for occupants.

Q426. Answer: B

Explanation:

A deluge sprinkler system releases water through all open nozzles at once upon detection of fire. It is designed for high-hazard areas where rapid fire spread is possible, such as chemical storage or aircraft hangars. This system provides maximum water application instantly, cooling combustible surfaces and controlling large fires quickly to limit damage and risk.

Q427. Answer: A

Explanation:

Class D fire extinguishers use specialized dry powders like sodium chloride or copper-based agents to smother burning metals. These powders absorb heat and form a crust that excludes oxygen, halting the reaction. They prevent violent flare-ups caused by contact with water or foam, making them essential for handling combustible metal fires safely and effectively.

Q428. Answer: A

Explanation:

Grounding and bonding eliminate static electricity buildup that could ignite flammable vapors or dust. By maintaining electrical continuity and directing static charges safely to ground, these measures prevent accidental discharges that can trigger fires or explosions in operations involving fuel transfer, solvent use, or powder handling.

Q429. Answer: D

Explanation:

Flame detectors identify infrared or ultraviolet radiation emitted by flames, providing immediate warning even before heat or smoke becomes significant. They are ideal for environments like fuel terminals or refineries where open flames can develop suddenly. Their sensitivity and speed make them critical for minimizing damage from rapid fire growth.

Q430. Answer: A

Explanation:

Fire compartmentation divides a building into sections with fire-resistant barriers to contain fire and smoke. This design slows fire progression, protects escape routes, and provides occupants more time to evacuate safely. It also helps limit property damage and supports firefighting operations by containing hazards to smaller areas.

Q431. Answer: B

Explanation:

A dry pipe system holds compressed air above a valve that prevents water from entering the pipes until activation. When a sprinkler opens, the air is released and water follows, discharging onto the fire. This setup is used in environments susceptible to freezing where water-filled systems could be damaged.

Q432. Answer: A

Explanation:

Maintaining cleanliness through regular housekeeping prevents accumulation of combustible dust that can ignite or explode. Proper ventilation, dust collection systems, and surface cleaning eliminate potential ignition sources and fuel layers. Preventing buildup in overhead beams, ducts, or machinery significantly reduces the risk of secondary explosions.

Q433. Answer: A

Explanation:

Fire load represents the total heat energy released if all combustibles in a space burn completely. It helps determine the required fire resistance rating, suppression system design, and evacuation planning. Assessing fire load ensures safety measures match potential hazards based on materials and occupancy types.

Q434. Answer: C

Explanation:

Fire dampers close automatically during a fire, preventing flames and smoke from moving through ductwork. They maintain the integrity of fire-rated barriers and protect adjacent areas. By containing hazards, these dampers help preserve safe egress routes and reduce smoke damage during building fires.

Q435. Answer: A

Explanation:

Oily rags can self-heat and ignite spontaneously due to oxidation. Storing them in covered, non-combustible containers prevents oxygen exposure and limits heat buildup. Regular disposal or laundering of rags further reduces risk, ensuring a clean and safe work environment free from fire hazards.

Q436. Answer: B

Explanation:

A fire-rated door assembly includes a door, frame, and hardware designed to withstand fire for a specified duration. It prevents flames and smoke from spreading between compartments while allowing people to exit safely. Proper installation and maintenance ensure that these doors function reliably during emergencies.

Q437. Answer: A

Explanation:

Automatic sprinklers contain a heat-sensitive element such as a glass bulb or fusible link that seals the water path until it reaches activation temperature. This ensures water is released only where heat from a fire is present, minimizing water damage and ensuring targeted suppression.

Q438. Answer: A

Explanation:

Properly illuminated exit routes guide occupants safely out of buildings during fires or power failures. Continuous lighting ensures visibility even in smoky or dark conditions, reducing panic and confusion. Clearly visible exit signs and lights help occupants find their way quickly to safety.

Q439. Answer: C

Explanation:

Improper storage of flammable liquids allows vapor accumulation that can ignite when exposed to heat or sparks. Storing containers in ventilated, fire-rated cabinets and maintaining proper labeling prevents ignition. Safe handling practices reduce the chance of explosions and protect workers and property from severe fire incidents.

Q440. Answer: A

Explanation:

Intumescent sealant expands when exposed to heat, sealing gaps and preventing fire and smoke from spreading through penetrations in walls or floors. It provides a flexible, long-lasting barrier that complements passive fire protection systems and helps maintain the overall integrity of fire-rated assemblies.

Q441. Answer: A

Explanation:

Clean agent systems use gaseous agents to suppress fire without moisture or residue, making them ideal for protecting electronics, archives, or control rooms. They extinguish fires rapidly while preserving sensitive materials and ensuring minimal downtime after activation, combining efficiency with property protection.

Q442. Answer: B

Explanation:

Clear access to fire hydrants ensures firefighters can immediately connect hoses for water supply during emergencies. Obstructions delay firefighting efforts, allowing fires to grow uncontrollably. Keeping a minimum clearance around hydrants guarantees fast, efficient fire response and improves overall emergency preparedness.

Q443. Answer: A

Explanation:

Carbon dioxide extinguishers displace oxygen and cool the fire without leaving conductive residue, making them safe for use on energized electrical equipment. They prevent short circuits and damage while quickly suppressing flames in computer rooms, panels, and other electrical systems.

Q444. Answer: A

Explanation:

Evacuation drills familiarize occupants with exit routes and emergency procedures, reducing confusion during real incidents. Regular practice ensures quick, organized responses that minimize injuries. Drills also allow organizations to identify bottlenecks, improve coordination, and refine their overall emergency response plans.

Q445. Answer: D

Explanation:

An automatic fire alarm system continuously monitors the environment for heat, smoke, or flame indicators. When activated, it alerts occupants and emergency personnel, allowing for rapid evacuation and fire control. This early warning capability significantly reduces risks to life and property.

Q446. Answer: A

Explanation:

Painting or obstructing sprinkler heads interferes with heat detection and water discharge patterns. These conditions delay activation or prevent effective coverage, compromising fire control. Regular inspection and unobstructed clearance ensure sprinklers respond promptly when needed.

Q447. Answer: A

Explanation:

Exit doors that open in the direction of travel enable people to leave swiftly without delay or crowding. This design reduces panic, prevents congestion, and ensures faster evacuation during emergencies. It is a fundamental principle in safe building egress planning.

Q448. Answer: C

Explanation:

Addressable fire alarm systems can pinpoint the exact location of the triggered detector or device. This precision allows faster investigation and response, minimizing damage and improving safety outcomes. The system's communication between devices enhances monitoring efficiency and fault detection.

Q449. Answer: A

Explanation:

Fire-resistant coatings, also called intumescent paints, swell when exposed to heat, forming an insulating char layer that protects steel from rapid temperature rise. This delay helps maintain structural integrity, allowing more time for evacuation and firefighting operations.

Q450. Answer: A

Explanation:

Fire drills test how well employees and emergency teams execute evacuation and response procedures. They identify weak points in coordination, communication, and timing, allowing continuous improvement. Conducting regular drills builds confidence, reduces panic, and ensures a prompt and orderly response during real emergencies.

Q451. Answer: B

Explanation:

The Comprehensive Environmental Response, Compensation, and Liability Act (CERCLA), known as Superfund, governs the cleanup of contaminated sites. It authorizes the federal government to respond directly to releases of hazardous substances and hold responsible parties liable for remediation costs. This law ensures contaminated land is cleaned safely and efficiently to protect public health and the environment.

Q452. Answer: C

Explanation:

The Environmental Protection Agency (EPA) enforces air quality standards through the Clean Air Act. It regulates emissions from stationary and mobile sources, setting National Ambient Air Quality Standards (NAAQS) to protect human health and the environment. The EPA oversees compliance, permitting, and enforcement actions for air pollution control programs nationwide.

Q453. Answer: B

Explanation:

An SPCC Plan prevents oil discharges into navigable waters and adjoining shorelines. Facilities storing significant quantities of oil must prepare and implement SPCC plans to prevent, contain, and mitigate spills. The plan includes inspection procedures, secondary containment, and emergency response steps, ensuring environmental protection and compliance with federal spill regulations.

Q454. Answer: A

Explanation:

P-list and U-list wastes under the Resource Conservation and Recovery Act (RCRA) identify specific commercial chemical products that are discarded or off-specification. These lists classify acute and toxic hazardous wastes, guiding proper management, labeling, and disposal to prevent environmental contamination and health hazards from unused or expired chemicals.

Q455. Answer: C

Explanation:

Nitrogen oxides (NO_x) react with volatile organic compounds (VOCs) in the presence of sunlight to form ozone and photochemical smog. This type of pollution causes respiratory irritation and environmental degradation. Controlling NO_x emissions from vehicles and industrial combustion sources is critical to reducing urban air pollution and improving visibility.

Q456. Answer: A

Explanation:

A cyclone separator removes particulate matter by creating a vortex that forces heavier particles to the walls of the chamber, where they fall into a collection bin. This mechanical separation process is efficient for large dust particles and serves as a pre-cleaning step before filtration or electrostatic precipitation in industrial air pollution control.

Q457. Answer: B

Explanation:

Secondary containment provides a protective barrier or enclosure capable of containing the full volume of a tank in the event of a spill or leak. It prevents hazardous materials from escaping into soil or waterways. Properly designed containment systems are essential for regulatory compliance and environmental protection during chemical storage and handling.

Q458. Answer: C

Explanation:

Aeration introduces oxygen into wastewater, promoting the growth of aerobic microorganisms that break down organic pollutants. This biological process significantly reduces the biological oxygen demand (BOD), improving effluent quality before discharge. Effective aeration ensures compliance with water quality standards and supports ecosystem health by preventing oxygen depletion in receiving waters.

Q459. Answer: B

Explanation:

The Resource Conservation and Recovery Act (RCRA) regulates hazardous waste from generation through transportation, treatment, storage, and disposal—known as cradle-to-grave management. It establishes permitting, manifesting, and compliance procedures that ensure waste is handled safely and responsibly to protect human health and the environment.

Q460. Answer: C

Explanation:

Sulfur dioxide (SO_2) released from burning fossil fuels reacts with atmospheric moisture to form sulfuric acid, contributing to acid rain. Acid deposition damages vegetation, aquatic ecosystems, and infrastructure. Controlling SO_2 emissions through scrubbers and low-sulfur fuels mitigates these harmful environmental effects and supports cleaner air quality.

Q461. Answer: D

Explanation:

The corrective action process ensures continuous improvement in an Environmental Management System by identifying, analyzing, and addressing nonconformities. Implementing corrective and preventive measures strengthens system performance, fosters accountability, and drives progress toward environmental objectives. This iterative improvement loop aligns with ISO 14001's "Plan-Do-Check-Act" framework for sustainable management practices.

Q462. Answer: B

Explanation:

Environmental Impact Assessments (EIAs) identify, predict, and evaluate potential environmental consequences of proposed projects before approval or implementation. They ensure that decision-makers consider environmental factors alongside technical and economic criteria. EIAs support sustainable development by preventing, minimizing, or mitigating adverse environmental impacts at the planning stage.

Q463. Answer: A

Explanation:

Incineration, or thermal oxidation, destroys volatile organic compounds (VOCs) by exposing them to high temperatures in the presence of oxygen. This process converts VOCs into carbon dioxide and water, effectively reducing air pollutant concentrations. Thermal oxidizers are widely used in manufacturing and coating industries to maintain air quality compliance and safety.

Q464. Answer: A

Explanation:

The Emergency Planning and Community Right-to-Know Act (EPCRA) enhances public access to information about chemical hazards in local communities. It requires facilities to report on storage, use, and releases of hazardous substances, ensuring emergency preparedness and transparency. The act empowers communities and first responders to plan effectively for chemical incidents.

Q465. Answer: B

Explanation:

A scrubber removes gaseous pollutants and particulates by bringing contaminated air into contact with a scrubbing liquid, typically water or a chemical reagent. This process dissolves or neutralizes pollutants before discharge. Scrubbers are essential for controlling emissions from combustion and industrial processes, improving air quality and regulatory compliance.

Q466. Answer: C

Explanation:

Nitrification is a two-step biological process that converts ammonia (NH_3) into nitrate (NO_3^-) via nitrite (NO_2^-). This oxidation process, carried out by nitrifying bacteria, is critical in secondary wastewater treatment to prevent ammonia toxicity and eutrophication in receiving waters. It supports balanced nutrient removal and environmental protection.

Q467. Answer: B

Explanation:

Sustainability programs aim to minimize environmental impact while maintaining operational efficiency and economic viability. They integrate resource conservation, waste reduction, and pollution prevention into daily operations. Sustainable practices improve long-term resilience, ensure regulatory compliance, and foster positive environmental and social outcomes for organizations.

Q468. Answer: C

Explanation:

The recovery phase focuses on restoring normal operations and repairing damage following an emergency or disaster. It involves rebuilding infrastructure, evaluating lessons learned, and implementing improvements to reduce future vulnerabilities. Recovery ensures community resilience and continuity of essential services after a disruptive environmental event.

Q469. Answer: B

Explanation:

Gravimetric sampling measures airborne particulate concentration by drawing air through a pre-weighed filter, collecting particles for later weighing. The mass difference indicates particulate concentration in milligrams per cubic meter. This method provides accurate quantitative data for occupational exposure assessment and environmental air monitoring programs.

Q470. Answer: C

Explanation:

The Toxic Substances Control Act (TSCA) regulates the manufacture, import, use, and disposal of chemical substances. It authorizes the evaluation of new and existing chemicals to ensure they do not pose unreasonable risks to human health or the environment. TSCA provides the framework for chemical inventory and reporting requirements in U.S. industry.

Q471. Answer: A

Explanation:

A Spill Contingency Plan details the specific steps and resources needed to respond to hazardous waste releases or spills. It includes notification, containment, cleanup, and recovery procedures to protect workers, communities, and the environment. Properly maintained contingency plans are essential for emergency preparedness and regulatory compliance.

Q472. Answer: B

Explanation:

Pollution prevention focuses on eliminating waste at the source by modifying processes, using non-toxic materials, and improving operational efficiency. Unlike end-of-pipe treatment, it prevents pollutants from being generated, reducing disposal costs and environmental impact. This proactive approach is a cornerstone of sustainable waste management and environmental responsibility.

Q473. Answer: B

Explanation:

The troposphere, extending from the Earth's surface up to about 10–15 kilometers, contains nearly all atmospheric water vapor and is where weather phenomena occur. It supports cloud formation, precipitation, and temperature variations that influence climate and air quality. Environmental monitoring focuses on this layer for pollution and meteorological analysis.

Q474. Answer: B

Explanation:

Non-point source pollution originates from diffuse sources such as agricultural runoff carrying fertilizers, pesticides, and sediments into water bodies. It is not discharged from a single, identifiable point, making control and regulation more challenging. Managing land use, buffer zones, and sustainable farming practices helps reduce non-point source pollution impacts.

Q475. Answer: C

Explanation:

Nitrate concentration indicates nutrient levels that contribute to eutrophication and algal blooms in aquatic systems. Elevated nitrates stimulate excessive plant growth, leading to oxygen depletion and ecosystem imbalance. Monitoring and managing nitrate inputs from agriculture and wastewater sources are critical for maintaining healthy water quality and preventing ecological damage.

Q476. Answer: C

Explanation:

The Clean Water Act establishes the framework for regulating discharges of pollutants into U.S. waters and setting quality standards for surface waters. It authorizes the National Pollutant Discharge Elimination System (NPDES) permitting program to control point source pollution. The act's primary goal is to restore and maintain the chemical, physical, and biological integrity of the nation's waters.

Q477. Answer: B

Explanation:

The Clean Air Act regulates air emissions from both stationary and mobile sources to protect public health and the environment. It authorizes the Environmental Protection Agency to establish and enforce National Ambient Air Quality Standards (NAAQS) for pollutants such as ozone, particulate matter, and sulfur dioxide. Its provisions drive air quality improvement nationwide.

Q478. Answer: B

Explanation:

Leachate is the liquid that forms when water percolates through waste materials, dissolving contaminants and carrying them into surrounding soils or groundwater. It is particularly associated with landfills and waste disposal sites. Proper leachate collection and treatment systems prevent groundwater pollution and protect surrounding ecosystems from contamination.

Q479. Answer: D

Explanation:

Corrective action review confirms that identified nonconformities have been addressed and that improvements align with regulatory and organizational requirements. It ensures compliance through systematic follow-up, demonstrating continuous improvement. Reviewing implemented actions validates the effectiveness of environmental management programs and reinforces accountability in maintaining compliance performance.

Q480. Answer: A

Explanation:

Composting decomposes organic waste through microbial activity, producing a nutrient-rich material used to improve soil quality. This natural process reduces landfill disposal, lowers greenhouse gas emissions, and supports sustainable waste management. Proper aeration, moisture control, and temperature regulation ensure effective composting and pathogen reduction.

Q481. Answer: B

Explanation:

Dissolved oxygen (DO) measures the amount of oxygen available in water for aquatic organisms. High DO levels indicate healthy water quality, while low DO suggests pollution or eutrophication. Maintaining adequate DO prevents fish kills and supports balanced aquatic ecosystems. Continuous monitoring helps ensure compliance with environmental water quality standards.

Q482. Answer: A

Explanation:

Recycling reintroduces recovered materials into the production cycle, conserving resources and reducing waste generation. It minimizes energy consumption and environmental impact compared to producing raw materials. Recycling programs enhance sustainability and compliance with waste management regulations while promoting circular economy practices.

Q483. Answer: C

Explanation:

The pH scale measures the acidity or alkalinity of water on a range from 0 to 14, where 7 is neutral. pH influences chemical reactions, biological activity, and metal solubility in aquatic environments. Maintaining balanced pH ensures stable water quality and effective treatment process control in environmental systems.

Q484. Answer: B

Explanation:

Flocculation aggregates fine suspended particles into larger clusters, or flocs, by gentle mixing and the addition of coagulants. These flocs are then removed during sedimentation or filtration. The process clarifies water, improving quality for potable or industrial use, and enhances the efficiency of downstream treatment stages.

Q485. Answer: C

Explanation:

The Maximum Contaminant Level (MCL) represents the highest allowable concentration of a contaminant in drinking water. It ensures safety for public consumption and is established under the Safe Drinking Water Act. MCLs protect against adverse health effects, ensuring consistent water quality and compliance with federal and state standards.

Q486. Answer: C

Explanation:

Source reduction—preventing waste generation at its origin—holds the highest environmental priority. It minimizes material use and pollution before they occur, conserving resources and reducing environmental impact. This proactive approach supports sustainability and aligns with pollution prevention principles central to effective environmental management systems.

Q487. Answer: B

Explanation:

Chlorofluorocarbons (CFCs) release chlorine atoms in the stratosphere, which catalytically destroy ozone molecules. This depletion reduces atmospheric protection against harmful ultraviolet radiation. Global agreements such as the Montreal Protocol have phased out CFC production to restore and preserve the ozone layer, enhancing long-term atmospheric stability and health protection.

Q488. Answer: B

Explanation:

Thermal pollution occurs when heated water from industrial cooling or energy production raises the temperature of nearby water bodies. Elevated temperatures reduce dissolved oxygen and disrupt aquatic ecosystems. Managing discharge temperatures through cooling systems and heat recovery technologies helps protect water quality and maintain ecological balance.

Q489. Answer: C

Explanation:

Renewable energy sources like solar, wind, and hydro reduce reliance on finite fossil fuels, providing long-term energy security and lower greenhouse gas emissions. Their sustainable nature supports carbon reduction goals and environmental protection, fostering clean energy transitions that balance ecological and economic benefits for future generations.

Q490. Answer: B

Explanation:

A composite liner combines a synthetic geomembrane with a compacted clay layer, creating a dual barrier that minimizes leachate migration. This design enhances protection against groundwater contamination and extends landfill lifespan. Proper liner construction and monitoring are critical for effective long-term waste containment and environmental compliance.

Q491. Answer: B

Explanation:

An electrostatic precipitator removes fine particulate matter by charging particles and collecting them on oppositely charged plates. This highly efficient device is used in power plants and industrial operations to control emissions. Its ability to handle large air volumes with minimal pressure drop makes it a cornerstone of air quality management.

Q492. Answer: B

Explanation:

The National Environmental Policy Act (NEPA) requires federal agencies to evaluate environmental impacts before undertaking major projects or decisions. It ensures that environmental factors are considered alongside technical and economic considerations. NEPA promotes transparency, public participation, and informed decision-making to prevent environmental degradation from government actions.

Q493. Answer: D

Explanation:

Acid rain forms when sulfur dioxide and nitrogen oxides react with water vapor to produce sulfuric and nitric acids. These acids fall to the ground as precipitation, damaging vegetation, aquatic habitats, and infrastructure. Reducing emissions from power plants and vehicles effectively mitigates acid rain formation and its widespread impacts.

Q494. Answer: B

Explanation:

Using low-emission or alternative fuel vehicles reduces air pollutant emissions such as carbon monoxide and nitrogen oxides. Fleet modernization, efficient route planning, and regular maintenance further decrease fuel consumption and greenhouse gas output. These practices enhance environmental performance and contribute to sustainable transportation operations.

Q495. Answer: B

Explanation:

Environmental risk assessment identifies hazards, evaluates exposure pathways, and estimates potential impacts on human health and ecosystems. It supports informed decision-making for managing chemical, biological, or physical risks. Conducting risk assessments ensures that mitigation strategies are proportional to potential harm, improving environmental protection and regulatory compliance.

Q496. Answer: A

Explanation:

Discharge refers to the intentional release of treated or controlled effluent into surface water bodies. It is regulated under discharge permits that ensure pollutants remain below acceptable limits. Properly managed discharges maintain ecological balance and compliance with environmental water quality standards to prevent contamination and protect aquatic life.

Q497. Answer: A

Explanation:

Universal waste includes common hazardous materials such as batteries, fluorescent lamps, and mercury-containing devices. This classification simplifies handling and recycling requirements while maintaining safety standards. Managing universal waste properly prevents environmental contamination and promotes efficient recycling or reclamation processes.

Q498. Answer: B

Explanation:

Wind speed and atmospheric stability determine how pollutants disperse in the atmosphere. Higher wind speeds enhance dilution, while stable conditions trap pollutants near the surface. Understanding meteorological effects is crucial for air quality modeling, emission control planning, and maintaining compliance with ambient air quality standards.

Q499. Answer: B

Explanation:

Incineration burns waste at high temperatures to significantly reduce volume and sometimes recover energy. It converts solid waste into ash, flue gas, and heat. Properly designed incinerators use emission control systems to prevent air pollution and comply with environmental regulations, providing an effective waste volume reduction method.

Q500. Answer: C

Explanation:

ISO 14001 provides a systematic framework for managing environmental responsibilities and improving performance. It promotes continual improvement through planning, implementation, and review. Organizations adopting ISO 14001 enhance compliance, resource efficiency, and stakeholder confidence, achieving measurable progress in environmental management and sustainability goals.

Questions (501-600)

Q501. Which parameter is used to assess organic pollution in wastewater?

A. Conductivity
B. Biological oxygen demand
C. pH
D. Nitrate level

Q502. Which environmental management approach integrates sustainability into all business operations?

A. Reactive compliance
B. Environmental stewardship
C. End-of-pipe treatment
D. Cost avoidance

Q503. Which device is commonly used to measure stack gas emissions?

A. Pitot tube
B. Spectrometer
C. Continuous emission monitor
D. Hygrometer

Q504. Which environmental document defines emergency actions for accidental releases?

A. Emergency response plan
B. Environmental policy
C. Audit checklist
D. Sustainability report

Q505. Which method separates oil from water in industrial wastewater treatment?

A. Filtration
B. Gravity separation
C. Ion exchange
D. Flotation

Q506. Which global agreement targets the reduction of greenhouse gas emissions?

A. Basel Convention
B. Montreal Protocol
C. Rotterdam Convention
D. Kyoto Protocol

Q507. Which term describes the process of reusing treated wastewater for beneficial purposes?

A. Reclamation
B. Infiltration
C. Diffusion
D. Purification

Q508. What is the main goal of an environmental audit?

A. Identify training needs
B. Verify compliance and performance improvement
C. Develop marketing plans
D. Increase production efficiency

Q509. Which waste disposal method is least preferable in the waste management hierarchy?

A. Recycling
B. Source reduction
C. Treatment
D. Landfilling

Q510. Which environmental sampling method measures contaminants in soil gas?

A. Passive diffusion sampling

B. Core drilling
C. Grab sampling
D. Composite sampling

Q511. Which pollutant is a major contributor to global warming?
A. Methane
B. Nitrogen
C. Sulfur dioxide
D. Carbon monoxide

Q512. What is the main function of a hazardous waste manifest?
A. Record workplace accidents
B. Determine waste toxicity levels
C. Track waste movement from generator to disposal
D. Document recycling progress

Q513. Which environmental control technology removes acidic gases from exhaust streams?
A. Absorber scrubber
B. Cyclone separator
C. Electrostatic precipitator
D. Baghouse filter

Q514. Which practice promotes energy conservation in industrial operations?
A. Continuous lighting
B. Heat recovery systems
C. Increased idle equipment
D. Manual process control

Q515. Which term describes the complete natural degradation of a material by microorganisms?
A. Photodegradation
B. Biodegradation
C. Oxidation
D. Absorption

Q516. What is the primary function of sedimentation in water treatment?
A. Remove dissolved gases
B. Allow solids to settle out of suspension
C. Neutralize acids
D. Eliminate pathogens

Q517. Which solid waste management strategy converts waste into useful energy?
A. Recycling
B. Incubation
C. Composting
D. Waste-to-energy

Q518. Which aspect of environmental management ensures readiness for natural disasters?
A. Maintenance program
B. Emergency preparedness planning
C. Recycling initiative
D. Emission inventory

Q519. Which pollutant causes eutrophication in surface waters?
A. Carbon monoxide
B. Nitrogen and phosphorus compounds
C. Lead
D. Ozone

Q520. Which global initiative promotes sustainable development through shared international goals?
A. Agenda 21

B. Clean Air Act
C. ISO 45001
D. CERCLA

Q521. Which air pollutant is directly measured to assess vehicle emission control effectiveness?
A. Nitrogen
B. Carbon monoxide
C. Methane
D. Argon

Q522. Which waste category includes spent solvents and paint residues?
A. F-listed waste
B. K-listed waste
C. P-listed waste
D. U-listed waste

Q523. What is the purpose of environmental benchmarking?
A. Develop advertising strategies
B. Identify profit margins
C. Compare an organization's environmental performance to best practices
D. Eliminate documentation requirements

Q524. Which practice minimizes environmental impact from chemical handling operations?
A. Using larger containers
B. Implementing spill prevention controls
C. Increasing chemical inventory
D. Using open transfer systems

Q525. What is the main benefit of conducting life cycle assessments (LCA)?
A. Increase production output
B. Evaluate environmental impacts of a product from creation to disposal
C. Determine marketing potential
D. Reduce maintenance schedules

Q526. Which principle best defines the goal of industrial hygiene?
A. To eliminate all workplace hazards
B. To anticipate, recognize, evaluate, and control hazards
C. To enforce workplace regulations
D. To monitor employee attendance

Q527. Which term describes the average airborne exposure level over a normal work shift?
A. Short-Term Exposure Limit (STEL)
B. Time-Weighted Average (TWA)
C. Ceiling Limit
D. Immediately Dangerous to Life or Health (IDLH)

Q528. What is the most effective method in the hierarchy of controls?
A. Personal protective equipment
B. Administrative controls
C. Substitution
D. Elimination

Q529. Which route of exposure allows chemicals to enter the body through the skin?
A. Inhalation
B. Absorption
C. Ingestion
D. Injection

Q530. Which control strategy reduces exposure by redesigning equipment or processes?
A. Administrative controls

B. Engineering controls
C. PPE
D. Medical surveillance

Q531. Which measurement represents sound intensity?
A. Decibel (dB)
B. Hertz (Hz)
C. Newton (N)
D. Pascal (Pa)

Q532. Which type of radiation consists of high-energy particles capable of ionizing matter?
A. Non-ionizing radiation
B. Ultraviolet radiation
C. Ionizing radiation
D. Infrared radiation

Q533. Which type of ventilation system captures contaminants at their source?
A. General dilution ventilation
B. Local exhaust ventilation
C. Natural ventilation
D. Positive pressure system

Q534. Which device is used to measure airborne dust concentration?
A. Noise dosimeter
B. Rotameter
C. Aerosol photometer
D. Anemometer

Q535. Which toxic effect occurs from a single high-level exposure?
A. Chronic effect
B. Acute effect
C. Subchronic effect
D. Latent effect

Q536. Which device measures sound levels in the workplace?
A. Sound level meter
B. Psychrometer
C. Gas detector
D. Barometer

Q537. Which chemical hazard is characterized by the ability to cause cancer?
A. Irritant
B. Sensitizer
C. Carcinogen
D. Corrosive

Q538. Which occupational exposure limit is based on a 15-minute period?
A. Ceiling limit
B. TWA
C. STEL
D. IDLH

Q539. Which type of respirator supplies air from a clean source?
A. Air-purifying respirator
B. Air-supplying respirator
C. Filtering facepiece
D. Chemical cartridge respirator

Q540. Which factor primarily influences chemical toxicity?
A. Dose

B. Temperature
C. Odor
D. Color
Q541. Which sampling method collects air contaminants over an extended period?
A. Grab sampling
B. Continuous monitoring
C. Integrated sampling
D. Spot sampling
Q542. What is the main function of a dosimeter in industrial hygiene?
A. To measure cumulative exposure over time
B. To filter air contaminants
C. To store calibration data
D. To analyze dust particle size
Q543. Which health effect results from prolonged repeated exposure to low levels of a toxic substance?
A. Thermal injury
B. Acute effect
C. Reversible irritation
D. Chronic effect
Q544. Which principle describes the relationship between radiation intensity and distance?
A. Inverse square law
B. Boyle's law
C. Dalton's law
D. Charles' law
Q545. Which control method limits worker exposure time to hazardous environments?
A. Administrative controls
B. Engineering controls
C. PPE
D. Ventilation
Q546. Which type of hazard is associated with excessive vibration exposure?
A. Biological hazard
B. Chemical hazard
C. Physical hazard
D. Ergonomic hazard
Q547. Which process removes contaminants by passing air through a porous medium?
A. Filtration
B. Absorption
C. Adsorption
D. Precipitation
Q548. Which biological hazard control practice protects against exposure to infectious materials?
A. Universal precautions
B. Handwashing only
C. Annual vaccination
D. Chemical neutralization
Q549. Which factor determines whether a noise source is harmful?
A. Frequency alone
B. Intensity and duration
C. Sound color
D. Direction of source
Q550. Which term defines the time required for half of a radioactive substance to decay?

A. Half-life
B. Decay constant
C. Exposure rate
D. Radiation flux
Q551. What is the main purpose of a hearing conservation program?
A. To prevent noise-induced hearing loss among employees
B. To monitor air quality
C. To control vibration hazards
D. To provide medical fitness exams
Q552. What type of respirator filter is required for protection against airborne biological agents?
A. P100 filter
B. Carbon filter
C. N95 filter
D. Gas cartridge filter
Q553. Which organ is primarily affected by exposure to hepatotoxic chemicals?
A. Liver
B. Kidney
C. Lung
D. Heart
Q554. What is the primary purpose of medical surveillance in occupational health programs?
A. To detect early signs of work-related illness or exposure
B. To monitor absenteeism rates
C. To evaluate production efficiency
D. To determine insurance premiums
Q555. Which type of sampling provides instantaneous measurement of airborne contaminants?
A. Time-weighted sampling
B. Integrated sampling
C. Passive diffusion sampling
D. Grab sampling
Q556. Which type of toxic agent primarily damages the kidneys?
A. Nephrotoxic agent
B. Neurotoxic agent
C. Pulmonary irritant
D. Hemotoxic agent
Q557. What is the primary function of a fume hood in laboratory settings?
A. To store chemicals
B. To remove hazardous vapors and protect users
C. To measure airflow
D. To sterilize glassware
Q558. Which physical hazard can lead to heat stress?
A. High ambient temperature and humidity
B. Poor lighting
C. Elevated noise levels
D. Cold air drafts
Q559. Which chemical property indicates how easily a substance evaporates into the air?
A. Vapor pressure
B. Flash point
C. pH
D. Solubility

Q560. Which system is primarily used for classifying chemical hazards in labels and safety data sheets?
A. Chemical Abstract System
B. OSHA Noise Standard
C. Globally Harmonized System (GHS)
D. EPA Waste Management Standard
Q561. Which term describes the movement of chemicals through the air?
A. Dispersion
B. Adsorption
C. Filtration
D. Neutralization
Q562. What is the function of an audiometric baseline test?
A. To analyze visual acuity
B. To determine blood pressure trends
C. To measure lung capacity
D. To establish initial hearing ability for future comparison
Q563. Which occupational disease is caused by inhalation of silica dust?
A. Silicosis
B. Asbestosis
C. Dermatitis
D. Pneumonia
Q564. Which control method replaces a hazardous material with a less hazardous one?
A. Substitution
B. Isolation
C. Ventilation
D. PPE
Q565. Which measurement unit is used to express noise exposure over time?
A. dBA
B. ppm
C. mg/m^3
D. cpm
Q566. Which principle defines the concept that "the dose makes the poison"?
A. Epidemiology
B. Toxicology
C. Pathology
D. Biostatistics
Q567. Which respiratory device supplies oxygen in oxygen-deficient environments?
A. Self-contained breathing apparatus (SCBA)
B. N95 respirator
C. Half-mask respirator
D. Dust mask
Q568. Which environmental factor is measured using a wet bulb globe thermometer?
A. Radiation
B. Vibration
C. Heat stress
D. Sound
Q569. Which hazard is associated with excessive ultraviolet radiation exposure?
A. Skin cancer and eye damage
B. Respiratory irritation
C. Hearing loss

D. Chemical burns
Q570. Which biological monitoring test measures exposure to lead?
A. Blood lead level test
B. Spirometry
C. Audiogram
D. Chest X-ray
Q571. Which workplace condition contributes most to musculoskeletal disorders?
A. Cold temperatures
B. High noise levels
C. Ultraviolet radiation
D. Repetitive motion and poor ergonomics
Q572. Which hazard is measured using a sound dosimeter?
A. Noise exposure
B. Airborne dust
C. Radiation levels
D. Heat index
Q573. What does PPE stand for in industrial hygiene?
A. Personal Protective Equipment
B. Physical Protection Element
C. Personal Pressure Evaluation
D. Personnel Protection Estimator
Q574. Which process removes hazardous vapors by drawing air through a carbon filter?
A. Filtration
B. Absorption
C. Adsorption
D. Diffusion
Q575. Which biological hazard is transmitted through contaminated needles or blood contact?
A. Hepatitis B virus
B. Legionella bacteria
C. Anthrax spores
D. Histoplasma fungus
Q576. Which part of the body is most affected by exposure to neurotoxic chemicals?
A. Nervous system
B. Respiratory system
C. Circulatory system
D. Digestive system
Q577. Which engineering control is used to prevent the spread of airborne contaminants in a workspace?
A. Administrative scheduling
B. Local exhaust ventilation
C. PPE rotation
D. Health monitoring
Q578. Which factor determines the likelihood of chemical absorption through the skin?
A. Chemical solubility and skin condition
B. Air temperature
C. Body weight
D. Noise level
Q579. What is the key goal of exposure assessment in industrial hygiene?
A. To manage staffing levels
B. To design production equipment

C. To calculate profit margins
D. To evaluate potential worker exposure to hazards
Q580. Which gas detection method continuously monitors the environment for hazardous vapors?
A. Direct-reading instrument
B. Grab sampling
C. Diffusion tube
D. Charcoal tube sampling
Q581. Which control measure focuses on reducing the number of exposed workers?
A. Administrative control
B. Engineering control
C. PPE
D. Substitution
Q582. Which factor influences the toxicity of a chemical substance?
A. Concentration, exposure duration, and route of entry
B. Color and odor
C. Packaging material
D. Temperature of storage
Q583. Which respiratory hazard results from insufficient oxygen in the air?
A. Asthma
B. Hypothermia
C. Pulmonary edema
D. Asphyxiation
Q584. Which device measures the flow rate of air during sampling?
A. Rotameter
B. Thermometer
C. Manometer
D. Barometer
Q585. Which physical hazard is associated with prolonged exposure to excessive noise?
A. Hearing loss
B. Skin burns
C. Heat exhaustion
D. Eye irritation
Q586. What is the term for the point where exposure levels exceed safety limits?
A. Threshold limit value (TLV)
B. Ceiling limit
C. Exposure index
D. Reference dose
Q587. Which chemical property is used to evaluate vapor hazards in confined spaces?
A. Vapor density
B. Specific gravity
C. Flash point
D. Solubility
Q588. Which device measures a worker's cumulative exposure to radiation?
A. Rotameter
B. Hygrometer
C. Dosimeter
D. Manometer
Q589. What is the best way to reduce hand-arm vibration syndrome risk?
A. Use anti-vibration tools and gloves
B. Increase work shifts

C. Raise air temperature
D. Lower lighting intensity
Q590. Which hazard control involves enclosing a process to isolate workers from contaminants?
A. Engineering control
B. Administrative control
C. PPE
D. Medical monitoring
Q591. Which type of sampling evaluates short-term fluctuations in airborne contaminant levels?
A. Grab sampling
B. Integrated sampling
C. Passive sampling
D. Continuous sampling
Q592. Which term describes the biological response that occurs only after prolonged exposure?
A. Chronic effect
B. Acute effect
C. Latent irritation
D. Temporary response
Q593. Which personal monitoring method evaluates noise exposure for an individual?
A. Sound level meter
B. Noise dosimeter
C. Microphone array
D. Acoustic calibrator
Q594. Which ventilation system replaces contaminated air with clean air in a workspace?
A. Dilution ventilation
B. Local exhaust ventilation
C. Isolation enclosure
D. Pressure hood system
Q595. Which measure prevents excessive dust accumulation in the workplace?
A. Installation of warning signs
B. Use of earplugs
C. Housekeeping and wet cleaning
D. Wearing flame-resistant clothing
Q596. Which occupational disease is associated with chronic exposure to asbestos fibers?
A. Asbestosis
B. Dermatitis
C. Silicosis
D. Tuberculosis
Q597. What is the primary purpose of biological monitoring?
A. To measure chemicals absorbed by the body
B. To evaluate sound exposure
C. To assess air velocity
D. To detect mechanical vibration
Q598. Which control technique limits worker access to hazardous areas?
A. Substitution
B. Engineering control
C. PPE
D. Administrative control
Q599. Which parameter describes the total sound energy exposure over time?
A. Noise dose
B. Sound intensity

C. Sound frequency
D. Pitch level

Q600. What is the main purpose of fit testing in respiratory protection programs?

A. To ensure a proper seal between the respirator and user's face
B. To measure lung capacity
C. To test oxygen levels
D. To check airflow in ducts

Answer keys and Explanations (501-600)

Q501. Answer: B

Explanation: Biological Oxygen Demand (BOD) measures the amount of oxygen microorganisms require to decompose organic matter in water. A high BOD indicates heavy organic pollution and poor water quality. It is a critical indicator for evaluating treatment efficiency and ensuring effluent meets discharge standards to protect aquatic life and environmental health.

Q502. Answer: B

Explanation: Environmental stewardship involves proactively managing resources and impacts across all business activities to promote sustainability. It extends beyond regulatory compliance to include conservation, waste reduction, and continuous improvement. This approach enhances corporate responsibility, reduces environmental footprint, and aligns organizational objectives with sustainable development goals.

Q503. Answer: C

Explanation: A Continuous Emission Monitor (CEM) continuously measures pollutants such as sulfur dioxide, nitrogen oxides, and particulates in exhaust gases. It provides real-time data for compliance and performance tracking. CEM systems are essential for industrial facilities under regulatory monitoring requirements, ensuring emission limits are consistently maintained.

Q504. Answer: A

Explanation: An emergency response plan outlines specific procedures to manage accidental releases or environmental incidents safely and efficiently. It includes notification, containment, cleanup, and communication protocols. Effective planning ensures worker safety, regulatory compliance, and minimized environmental damage during emergencies.

Q505. Answer: B

Explanation: Gravity separation removes oil from water by exploiting density differences between oil and water phases. Lighter oil rises to the surface for collection, while heavier solids settle. This simple, effective process is often used as a primary treatment step in industrial wastewater systems to prevent pollution and improve downstream treatment efficiency.

Q506. Answer: D

Explanation: The Kyoto Protocol is an international treaty that commits participating countries to reduce greenhouse gas emissions based on agreed targets. It promotes carbon accounting and mechanisms like emissions trading. The agreement fosters international cooperation to mitigate climate change and supports the transition toward cleaner, low-carbon energy sources.

Q507. Answer: A

Explanation: Reclamation refers to treating wastewater to a quality suitable for reuse, such as irrigation, industrial processes, or groundwater recharge. This practice conserves freshwater resources, reduces discharge volumes, and supports sustainable water management. Reclaimed water plays a vital role in addressing water scarcity and promoting circular resource use.

Q508. Answer: B

Explanation: Environmental audits systematically assess compliance with laws, regulations, and internal policies. They identify gaps, evaluate control effectiveness, and recommend improvements. The goal is to enhance environmental performance and reduce risks by ensuring that environmental management systems operate efficiently and align with established objectives.

Q509. Answer: D

Explanation: Landfilling is the least preferred waste disposal method because it consumes land, produces leachate, and generates methane emissions. Modern environmental management prioritizes waste prevention, reuse, and recycling to minimize landfill dependence. When landfilling is unavoidable, engineering controls mitigate environmental impacts through containment and gas collection systems.

Q510. Answer: A

Explanation: Passive diffusion sampling collects soil gas contaminants by allowing vapors to diffuse into a sorbent medium over time. This non-intrusive technique provides representative data on volatile compounds for site assessment and remediation planning. It is widely used to monitor vapor intrusion and evaluate subsurface contamination.

Q511. Answer: A

Explanation: Methane is a potent greenhouse gas with a global warming potential many times higher than carbon dioxide. It originates from agriculture, waste decomposition, and fossil fuel operations. Reducing methane emissions is critical for mitigating climate change and improving overall air quality.

Q512. Answer: C

Explanation: A hazardous waste manifest documents the transportation of hazardous waste from its point of generation to final disposal. This tracking system ensures accountability, proper handling, and regulatory compliance throughout the waste lifecycle. It prevents illegal dumping and supports transparent waste management practices.

Q513. Answer: A

Explanation: An absorber scrubber removes acidic gases such as sulfur dioxide and hydrogen chloride by contacting them with an alkaline solution. The neutralization reaction converts harmful gases into benign compounds. This technology is essential in power plants and chemical industries for air pollution control and compliance.

Q514. Answer: B

Explanation: Heat recovery systems capture waste heat from industrial processes and reuse it for preheating or power generation. This improves energy efficiency, reduces operating costs, and lowers greenhouse gas emissions. Implementing energy recovery supports sustainability and enhances overall process performance.

Q515. Answer: B

Explanation: Biodegradation occurs when microorganisms such as bacteria and fungi break down organic substances into simpler, non-toxic compounds like water and carbon dioxide. This natural process is crucial for waste treatment and environmental remediation, supporting ecological balance and pollution reduction.

Q516. Answer: B

Explanation: Sedimentation allows suspended solids in water to settle by gravity, producing clarified water for subsequent filtration or disinfection. It reduces turbidity and load on downstream treatment processes, improving overall efficiency. Effective sedimentation is essential for both municipal and industrial water treatment systems.

Q517. Answer: D

Explanation: Waste-to-energy processes convert municipal solid waste into electricity or heat through combustion or gasification. This reduces landfill use and provides a renewable energy source. Properly managed systems minimize emissions and support sustainable waste management by integrating energy recovery with disposal.

Q518. Answer: B

Explanation: Emergency preparedness planning ensures that organizations are ready to respond effectively to natural or technological disasters. It includes risk assessment, training, and coordination to minimize environmental and operational impacts. Proper planning supports resilience, continuity, and rapid recovery following emergencies.

Q519. Answer: B

Explanation: Nitrogen and phosphorus compounds promote excessive algal growth in water bodies, a process known as eutrophication. When algae decompose, oxygen levels drop, harming aquatic life. Controlling nutrient inputs from agricultural runoff and wastewater discharges prevents eutrophication and preserves water quality.

Q520. Answer: A

Explanation: Agenda 21 is a global sustainability framework established at the 1992 Earth Summit. It provides guidelines for integrating environmental, economic, and social objectives into national and local policies. Agenda 21 encourages sustainable resource management and community participation to achieve long-term global development goals.

Q521. Answer: B

Explanation: Carbon monoxide is a key indicator pollutant for evaluating vehicle emission control efficiency. It forms due to incomplete fuel combustion and is monitored to verify the performance of catalytic converters. Reducing CO emissions improves air quality and aligns with transportation emission standards.

Q522. Answer: A

Explanation: F-listed wastes represent common industrial hazardous wastes such as spent solvents and paint residues. They originate from non-specific sources and are regulated for safe handling and disposal. Managing F-listed waste prevents soil and groundwater contamination while ensuring compliance with hazardous waste regulations.

Q523. Answer: C

Explanation: Environmental benchmarking compares an organization’s performance metrics against industry best practices or standards. It helps identify improvement opportunities and encourages adoption of efficient technologies and methods. Benchmarking supports continuous improvement and competitive environmental performance.

Q524. Answer: B

Explanation: Spill prevention controls, such as secondary containment and proper transfer systems, minimize environmental impact by preventing chemical releases. Implementing regular inspections and employee training enhances control effectiveness. These proactive measures protect ecosystems, reduce cleanup costs, and maintain compliance with environmental safety requirements.

Q525. Answer: B

Explanation: Life Cycle Assessment (LCA) evaluates a product's environmental impacts across all stages—from raw material extraction to disposal. This holistic approach identifies opportunities to reduce emissions, energy use, and waste generation. LCA supports sustainable product design and informed decision-making for environmental improvement.

Q526. Answer: B

Explanation: The purpose of industrial hygiene is to anticipate, recognize, evaluate, and control environmental factors or stresses that may cause illness, impaired health, or discomfort among workers. This proactive approach ensures hazards are identified before exposure occurs, allowing the application of engineering, administrative, and personal protective controls to protect worker health and safety effectively.

Q527. Answer: B

Explanation: The Time-Weighted Average (TWA) represents the average exposure level to a contaminant over an eight-hour workday or forty-hour workweek. It reflects cumulative exposure rather than instantaneous peaks, helping determine compliance with occupational exposure limits. Maintaining exposure below the TWA minimizes chronic health effects and ensures a safe long-term work environment.

Q528. Answer: D

Explanation: Elimination, the highest level in the hierarchy of controls, removes the hazard entirely from the workplace. This approach prevents exposure by addressing risks at the source. Once the hazard no longer exists, no further control is needed, making elimination the most reliable and permanent means of ensuring occupational health and safety.

Q529. Answer: B

Explanation: Absorption occurs when chemicals penetrate the skin and enter the bloodstream, often without immediate symptoms. Many solvents, pesticides, and metals can pass through the skin, causing systemic toxicity. Proper protective clothing, gloves, and hygiene practices prevent skin absorption and protect workers from harmful chemical exposure during occupational activities.

Q530. Answer: B

Explanation: Engineering controls isolate or remove hazards through physical modifications, such as improved ventilation, containment, or equipment redesign. These controls act at the source of exposure and are more effective than administrative methods or PPE. They protect all workers equally and reduce dependence on individual behavior for hazard prevention.

Q531. Answer: A

Explanation: Sound intensity is measured in decibels (dB), a logarithmic scale representing the ratio between a sound's pressure and a reference level. The decibel scale expresses large variations in intensity compactly, allowing accurate evaluation of workplace noise exposure and determination of hearing protection requirements based on noise level assessments.

Q532. Answer: C

Explanation: Ionizing radiation carries sufficient energy to remove tightly bound electrons from atoms, creating ions. Examples include alpha, beta, and gamma radiation, which can damage living tissue and cause cellular mutations. Proper shielding, distance, and exposure time control are critical in minimizing health risks associated with ionizing radiation.

Q533. Answer: B

Explanation: Local exhaust ventilation (LEV) removes contaminants directly from the point of generation before they disperse into the workplace air. It uses hoods, ducts, and filters to capture dust, fumes, and vapors efficiently. This method protects workers by controlling hazards at their origin, reducing airborne exposure concentrations effectively.

Q534. Answer: C

Explanation: An aerosol photometer measures airborne particulate concentration by detecting light scattered by dust particles in a sampled air stream. It provides real-time readings, allowing hygienists to monitor air quality and verify effectiveness of ventilation systems or control measures in maintaining safe respirable dust levels.

Q535. Answer: B

Explanation: An acute effect results from one-time exposure to a high concentration of a hazardous substance, causing immediate or short-term symptoms like dizziness or burns. Unlike chronic effects, which develop from long-term exposure, acute effects appear quickly and require rapid medical attention to prevent lasting harm.

Q536. Answer: A

Explanation: A sound level meter quantifies noise intensity using a microphone and electronic circuitry to measure sound pressure levels in decibels. These readings help identify hazardous noise areas and determine the need for hearing protection programs to prevent occupational hearing loss among workers in high-noise environments.

Q537. Answer: C

Explanation: A carcinogen is a substance capable of inducing cancer by altering cellular DNA and promoting uncontrolled cell growth. Prolonged exposure to carcinogens, even at low concentrations, can lead to tumor development. Controlling such exposures through substitution, engineering measures, and medical surveillance safeguards worker health effectively.

Q538. Answer: C

Explanation: The Short-Term Exposure Limit (STEL) defines the maximum concentration of a substance workers can be exposed to for up to 15 minutes without experiencing adverse effects. It prevents short-duration overexposures that could cause irritation or systemic harm, ensuring short-term exposure peaks remain within safe boundaries.

Q539. Answer: B

Explanation: An air-supplying respirator provides breathing air from an uncontaminated source such as a compressed air tank or airline system. It ensures consistent oxygen availability in environments with toxic atmospheres or low oxygen levels. This type offers superior protection compared to filtering devices that rely on ambient air purification.

Q540. Answer: A

Explanation: The toxic effect of any substance depends on its dose, or the amount entering the body over time. The relationship between dose and response is a key principle of toxicology, emphasizing that even benign substances can become harmful at high enough concentrations, while hazardous substances may be tolerated in small doses.

Q541. Answer: C

Explanation: Integrated sampling collects airborne contaminants over several hours, producing an average concentration value that reflects long-term exposure conditions. Air passes through filters or sorbent tubes, which trap contaminants for laboratory analysis. This approach helps assess compliance with exposure limits and identify chronic exposure risks accurately.

Q542. Answer: A

Explanation: A dosimeter records cumulative exposure to hazards such as radiation or noise over a defined period. It provides data reflecting total worker exposure, allowing accurate evaluation against permissible limits. This personal monitoring tool ensures control measures effectively protect workers from overexposure in various occupational settings.

Q543. Answer: D

Explanation: Chronic effects develop gradually due to long-term exposure to low concentrations of hazardous substances. Symptoms may appear after months or years, affecting organs like the liver, kidneys, or lungs. Continuous monitoring and control of chronic exposures prevent irreversible health damage and promote sustainable workplace safety.

Q544. Answer: A

Explanation: The inverse square law states that radiation intensity decreases proportionally to the square of the distance from the source. Doubling the distance from a radiation source reduces exposure by a factor of four. Maintaining safe distance is an effective way to control radiation exposure without additional shielding.

Q545. Answer: A

Explanation: Administrative controls manage worker behavior and schedule to reduce exposure, such as rotating shifts, restricting access, or enforcing rest breaks. They minimize the time workers spend in hazardous environments, lowering risk when elimination or engineering controls cannot fully remove exposure.

Q546. Answer: C

Explanation: Excessive vibration is a physical hazard that can cause circulatory disorders, nerve damage, and musculoskeletal stress. Continuous exposure to vibrating tools or equipment leads to hand-arm vibration syndrome or lower back pain. Controlling vibration through equipment design, maintenance, and use of protective gloves protects worker health.

Q547. Answer: A

Explanation: Filtration separates particles from air by forcing it through a porous medium like a filter. Solid contaminants are trapped while clean air passes through. This process is used in respirators and ventilation systems to control dust and aerosols, improving air quality and worker safety.

Q548. Answer: A

Explanation: Universal precautions involve treating all human blood and certain bodily fluids as potentially infectious, regardless of perceived risk. Using gloves, masks, and protective clothing prevents transmission of diseases like hepatitis and HIV. These practices form the foundation of infection control programs in healthcare and laboratory settings.

Q549. Answer: B

Explanation: Noise hazard potential depends on both intensity (sound level) and duration of exposure. High-intensity sounds cause immediate hearing damage, while prolonged moderate noise leads to gradual loss. Controlling either factor—by reducing volume or limiting exposure time—effectively protects hearing and ensures compliance with occupational noise limits.

Q550. Answer: A

Explanation: Half-life is the period needed for half of the atoms in a radioactive material to decay into a more stable form. It characterizes the rate of radioactive decay and helps determine safe handling, storage, and exposure control. Understanding half-life is essential for managing radiation safety effectively.

Q551. Answer: A

Explanation: A hearing conservation program aims to prevent occupational hearing loss by identifying hazardous noise levels, monitoring exposure, providing protective equipment, and conducting regular audiometric testing. It ensures that engineering and administrative controls are implemented to reduce noise at its source, preserving long-term employee hearing health and compliance with workplace safety standards.

Q552. Answer: C

Explanation: An N95 filter removes at least 95% of airborne particles, including biological agents such as bacteria and viruses. These respirators are widely used in healthcare and laboratory environments to prevent respiratory infections. The N95 designation ensures a high level of filtration efficiency and a tight face seal for effective protection.

Q553. Answer: A

Explanation: Hepatotoxic chemicals target the liver, impairing its ability to metabolize toxins and regulate vital biochemical functions. Substances such as carbon tetrachloride and certain industrial solvents can cause liver inflammation, necrosis, or cirrhosis. Controlling exposure through engineering controls and proper handling practices prevents long-term liver damage.

Q554. Answer: A

Explanation: Medical surveillance tracks employee health over time to identify early indicators of occupational illness or exposure before serious harm occurs. Regular examinations and biological monitoring help evaluate the effectiveness of workplace controls and ensure that protective measures remain sufficient to safeguard worker health.

Q555. Answer: D

Explanation: Grab sampling captures an air sample at a single point in time to measure contaminant concentrations instantly. It is useful for identifying peak exposures or investigating unexpected releases. Although it provides a snapshot rather than long-term data, it helps evaluate short-term exposure hazards effectively.

Q556. Answer: A

Explanation: Nephrotoxic agents cause kidney damage by disrupting filtration and waste removal processes. Heavy metals like lead, mercury, and cadmium are common nephrotoxins. Prolonged exposure can result in kidney failure, fluid imbalance, and systemic toxicity, emphasizing the need for exposure control and biological monitoring programs.

Q557. Answer: B

Explanation: A fume hood captures and exhausts hazardous vapors, gases, and particulates to protect laboratory workers from inhalation exposure. It maintains negative pressure to contain contaminants and prevent their release into the workspace. Proper use and maintenance of fume hoods are essential for effective exposure control.

Q558. Answer: A

Explanation: Exposure to high temperature and humidity causes the body to retain heat, leading to heat stress or heat stroke. Workers performing strenuous tasks in hot environments must have adequate hydration, rest breaks, and ventilation. Controlling ambient conditions minimizes physiological strain and prevents heat-related illnesses.

Q559. Answer: A

Explanation: Vapor pressure measures a substance's tendency to evaporate at a given temperature. Chemicals with high vapor pressure readily form vapors that can lead to inhalation hazards. Understanding vapor pressure helps assess potential airborne concentrations and determine appropriate ventilation or control measures for safe handling.

Q560. Answer: C

Explanation: The Globally Harmonized System (GHS) standardizes the classification and labeling of chemicals by defining hazard categories, pictograms, and precautionary statements. It ensures consistent communication of chemical hazards worldwide, helping employers and employees recognize and manage risks safely through standardized hazard information.

Q561. Answer: A

Explanation: Dispersion refers to the process of a chemical spreading through air due to wind, temperature, or turbulence. It determines how far and fast airborne contaminants travel, influencing exposure assessments and control zone design. Understanding dispersion helps predict contaminant behavior for effective hazard control planning.

Q562. Answer: D

Explanation: A baseline audiometric test records an employee's initial hearing ability before significant occupational noise exposure. Subsequent tests are compared to this baseline to detect any early hearing changes. Regular monitoring allows timely intervention and ensures the effectiveness of hearing protection programs.

Q563. Answer: A

Explanation: Silicosis is a lung disease resulting from prolonged inhalation of respirable crystalline silica dust. The particles cause inflammation and scarring of lung tissue, leading to reduced breathing capacity. Effective ventilation, dust suppression, and respiratory protection prevent this irreversible occupational disease.

Q564. Answer: A

Explanation: Substitution reduces risk by replacing a hazardous substance or process with a safer alternative that performs the same function. This proactive approach minimizes exposure at its source without relying on personal protective equipment. It's a key strategy within the hierarchy of hazard controls.

Q565. Answer: A

Explanation: Noise exposure is expressed in decibels A-weighted (dBA), reflecting how the human ear perceives sound frequencies. The dBA scale helps assess noise intensity over time and determine if hearing protection or control measures are required to prevent occupational hearing damage.

Q566. Answer: B

Explanation: Toxicology is the scientific study of harmful effects caused by chemical substances on living organisms. It is based on the principle that toxicity depends on the dose—every substance can be toxic at a high enough concentration. Understanding this helps assess risk and establish safe exposure limits.

Q567. Answer: A

Explanation: A self-contained breathing apparatus (SCBA) provides breathable air from a tank, making it essential in oxygen-deficient or highly toxic atmospheres. It ensures safety for workers entering confined spaces or responding to chemical releases where ambient air cannot support life.

Q568. Answer: C

Explanation: A wet bulb globe thermometer measures heat stress by combining temperature, humidity, air movement, and radiant heat effects. This composite reading determines how environmental conditions affect the body's heat balance and helps prevent heat-related illness by guiding safe work-rest cycles.

Q569. Answer: A

Explanation: Ultraviolet (UV) radiation damages skin cells and eyes, increasing the risk of skin cancer, sunburn, and cataracts. Workers exposed to sunlight or UV lamps require protective clothing, shields, and eyewear to reduce exposure and prevent long-term health complications.

Q570. Answer: A

Explanation: A blood lead level test quantifies lead absorbed into the bloodstream, indicating cumulative exposure. It helps evaluate the effectiveness of workplace controls and guides medical intervention when necessary. Regular testing ensures that workers remain within safe biological exposure limits.

Q571. Answer: D

Explanation: Repetitive motion, awkward posture, and poor workstation design cause musculoskeletal disorders by straining muscles and joints. Implementing ergonomic solutions, proper tool design, and task variation reduces stress on the body and prevents injury, improving worker comfort and productivity.

Q572. Answer: A

Explanation: A sound dosimeter measures cumulative noise exposure over a worker's shift. It records variations in sound levels and calculates total exposure to determine if hearing conservation measures are necessary. This personal monitoring tool ensures compliance with occupational noise limits.

Q573. Answer: A

Explanation: Personal Protective Equipment (PPE) includes items such as gloves, respirators, and safety goggles designed to protect workers from exposure to chemical, physical, or biological hazards. It is the last line of defense when engineering and administrative controls cannot fully eliminate risks.

Q574. Answer: C

Explanation: Adsorption occurs when gas molecules adhere to the surface of a solid material such as activated carbon. This process removes hazardous vapors from air streams efficiently, making it a key feature in respirators and air purification systems designed to protect workers from airborne toxins.

Q575. Answer: A

Explanation: Hepatitis B virus is a bloodborne pathogen transmitted through contact with infected blood or contaminated needles. Workers in healthcare or laboratory settings must follow standard precautions, vaccination programs, and proper sharps disposal to prevent occupational infection.

Q576. Answer: A

Explanation: Neurotoxic chemicals disrupt the normal function of the nervous system by damaging nerve cells or interfering with signal transmission. Symptoms may include numbness, tremors, or cognitive impairment. Common neurotoxins include solvents, pesticides, and heavy metals. Preventing exposure through proper ventilation, protective equipment, and substitution of safer chemicals preserves neurological health and performance.

Q577. Answer: B

Explanation: Local exhaust ventilation captures and removes airborne contaminants such as dust, vapors, or fumes at their source before they disperse into the work environment. It is one of the most effective engineering controls, ensuring cleaner air and protecting workers by maintaining exposure levels below permissible limits.

Q578. Answer: A

Explanation: The rate of dermal absorption depends on chemical solubility and the condition of the skin. Lipid-soluble compounds penetrate skin faster, especially if the skin is damaged or moist. Understanding these factors is critical for selecting appropriate gloves and protective clothing to minimize transdermal chemical exposure.

Q579. Answer: D

Explanation: Exposure assessment measures and evaluates the extent of worker exposure to hazardous agents. It involves air sampling, observation, and data analysis to determine compliance with exposure limits. The results guide control measures, ensuring that hazards are effectively mitigated and occupational exposure risks are minimized.

Q580. Answer: A

Explanation: Direct-reading instruments provide continuous or real-time detection of gases and vapors in the workplace. They instantly display concentration levels, allowing prompt response to leaks or hazardous releases. This continuous monitoring enhances worker protection and ensures immediate corrective actions are taken when thresholds are exceeded.

Q581. Answer: A

Explanation: Administrative controls minimize exposure by adjusting work schedules, limiting time in hazardous areas, and rotating staff. They reduce exposure duration and number of affected workers. Although less effective than engineering controls, they play an important role when elimination or substitution is not feasible.

Q582. Answer: A

Explanation: A chemical's toxicity is influenced by its concentration, duration of exposure, and route of entry—whether inhaled, ingested, or absorbed. Higher concentrations or longer exposure durations increase health risks. Understanding these factors helps industrial hygienists assess risk levels and implement effective exposure control strategies.

Q583. Answer: D

Explanation: Asphyxiation occurs when oxygen concentration in the air drops below safe levels, leading to unconsciousness or death. It may result from displacement by inert gases or confined space entry without ventilation. Monitoring oxygen levels and using supplied-air respirators prevent oxygen-deficient exposure incidents.

Q584. Answer: A

Explanation: A rotameter measures air flow rate in sampling equipment, ensuring proper calibration and accurate sample collection. Maintaining correct flow is vital for reliable data when assessing airborne contaminant concentrations and determining compliance with occupational exposure limits.

Q585. Answer: A

Explanation: Prolonged exposure to high noise levels damages the sensory cells in the inner ear, causing irreversible hearing loss. Continuous noise monitoring, engineering noise reduction, and use of hearing protection devices are crucial to prevent permanent auditory damage and preserve worker hearing health.

Q586. Answer: B

Explanation: A ceiling limit represents the maximum concentration of a substance that must not be exceeded at any time during exposure. It is critical for chemicals that cause immediate harm even at short durations. Adhering to ceiling limits ensures prompt protection against acute exposure effects.

Q587. Answer: A

Explanation: Vapor density compares the weight of a gas or vapor to air. Vapors heavier than air can accumulate in low-lying areas, creating confined space hazards. Recognizing vapor density helps in designing ventilation systems and assessing the need for atmospheric testing before entry.

Q588. Answer: C

Explanation: A dosimeter records cumulative radiation exposure over time, allowing comparison against safe exposure limits. It is typically worn on the body to monitor personal dose levels. This device ensures early detection of overexposure and supports compliance with occupational radiation protection requirements.

Q589. Answer: A

Explanation: Using anti-vibration tools and gloves minimizes vibration transmission to the hands and arms. Regular tool maintenance, limited exposure time, and rest breaks further reduce risk. Implementing these controls prevents circulatory and nerve disorders associated with vibration exposure.

Q590. Answer: A

Explanation: Engineering controls include process enclosure, which physically isolates workers from hazardous materials or emissions. By containing the source, exposure levels are reduced for everyone, providing a reliable and sustainable method of maintaining workplace air quality and safety.

Q591. Answer: D

Explanation: Continuous sampling provides real-time data on changing airborne contaminant concentrations. This method detects sudden spikes in exposure levels that may go unnoticed in integrated sampling. Continuous monitoring helps ensure that control systems respond promptly to protect workers from short-term overexposures.

Q592. Answer: A

Explanation: Chronic effects appear after long-term or repeated exposure to hazardous substances. They often involve slow, progressive damage to organs such as the liver, lungs, or kidneys. Continuous exposure monitoring and preventive measures protect against these delayed health impacts.

Q593. Answer: B

Explanation: A noise dosimeter measures cumulative noise exposure experienced by an individual worker throughout their shift. It records varying sound levels to calculate total dose, ensuring compliance with permissible exposure limits and verifying hearing protection effectiveness.

Q594. Answer: A

Explanation: Dilution ventilation reduces airborne contaminant concentrations by mixing clean air with contaminated air. It is suitable for low-toxicity vapors or gases dispersed evenly in large spaces. This system helps maintain air quality by controlling overall contaminant levels.

Q595. Answer: C

Explanation: Regular housekeeping, including wet cleaning or vacuum systems, prevents dust accumulation that could become airborne or combustible. Maintaining cleanliness reduces respiratory exposure and explosion risks, ensuring a safe and healthy work environment for employees.

Q596. Answer: A

Explanation: Asbestosis is a chronic lung condition caused by inhalation of asbestos fibers, leading to scarring and reduced lung function. Prolonged exposure can also increase cancer risk. Proper respiratory protection and removal procedures prevent exposure to airborne asbestos fibers in industrial settings.

Q597. Answer: A

Explanation: Biological monitoring determines the amount of chemical absorbed into a worker's body through analysis of blood, urine, or breath samples. It provides direct evidence of exposure and effectiveness of controls, ensuring worker safety through continuous health surveillance.

Q598. Answer: D

Explanation: Administrative controls limit worker access by restricting entry to hazardous zones, posting warning signs, and implementing permit systems. These measures minimize exposure opportunities and enhance safety when engineering solutions alone cannot fully eliminate hazards.

Q599. Answer: A

Explanation: Noise dose represents the total accumulated sound energy experienced by a worker during a specific period, typically measured by dosimeters. It helps evaluate whether the exposure exceeds permissible levels and guides corrective actions to maintain hearing conservation.

Q600. Answer: A

Explanation: Fit testing verifies that a respirator forms an adequate seal on the wearer's face, ensuring effective protection from airborne contaminants. It confirms compatibility, identifies leaks, and ensures that the respirator performs as intended under actual workplace conditions.

Questions (601-700)

Q601. Which method helps prevent respiratory diseases in dusty workplaces?

A. Increased humidity only

B. Effective ventilation and dust suppression

C. Use of ear protection

D. Wearing flame-resistant clothing

Q602. Which occupational hazard can cause hand-arm vibration syndrome?

A. Prolonged use of vibrating tools

B. Exposure to low temperatures

C. Contact with irritant chemicals

D. Excessive noise

Q603. Which principle describes how radiation exposure decreases with distance?

A. Inverse square law

B. Boyle's law

C. Dalton's law

D. Charles's law

Q604. What is the function of local exhaust ventilation systems?

A. To filter ambient air for comfort

B. To increase temperature in the workspace

C. To remove contaminants directly at their source

D. To provide cooling airflow only

Q605. Which factor most influences the rate of chemical absorption through skin?

A. Skin condition and chemical solubility

B. Air humidity

C. Lighting level

D. Air velocity

Q606. Which occupational disease is associated with exposure to high noise levels?

A. Dermatitis

B. Silicosis

C. Asbestosis

D. Noise-induced hearing loss

Q607. Which type of sampling collects contaminants continuously for long-term assessment?

A. Integrated sampling

B. Grab sampling

C. Spot sampling

D. Batch sampling

Q608. What is the primary function of an occupational health nurse?

A. To design process equipment

B. To monitor and promote worker health and well-being

C. To maintain building ventilation

D. To supervise fire suppression systems

Q609. Which measurement describes the average noise exposure over an 8-hour workday?

A. Time-weighted average (TWA)

B. Peak noise level

C. Frequency response

D. Instantaneous level

Q610. Which process uses sorbent tubes to collect airborne contaminants?

A. Active air sampling

B. Passive sampling

C. Continuous monitoring
D. Gas chromatography
Q611. Which physiological condition occurs when the body absorbs excessive heat?
A. Anemia
B. Hypothermia
C. Heat stroke
D. Asthma
Q612. Which personal protective equipment prevents respiratory hazards from airborne contaminants?
A. Respirator
B. Safety glasses
C. Hearing protectors
D. Apron
Q613. Which chemical hazard affects the blood's ability to carry oxygen?
A. Carbon monoxide
B. Sulfur dioxide
C. Ammonia
D. Methane
Q614. Which occupational exposure may lead to skin sensitization?
A. Exposure to helium
B. Exposure to carbon dioxide
C. Contact with nitrogen gas
D. Contact with epoxy resins
Q615. What is the main purpose of a hazard communication program?
A. To inform workers about chemical hazards and safe handling
B. To reduce insurance premiums
C. To train only supervisors
D. To control production flow
Q616. Which type of noise exposure is most harmful to hearing?
A. Low-frequency background noise
B. Continuous high-intensity noise
C. Short-duration soft sounds
D. Rhythmic ambient noise
Q617. Which process describes the transfer of toxic material from workplace air to the bloodstream through the lungs?
A. Inhalation absorption
B. Dermal contact
C. Injection
D. Ingestion
Q618. What is the function of a biological exposure index (BEI)?
A. To evaluate internal dose of a chemical absorbed by a worker
B. To measure ambient air levels
C. To assess room temperature
D. To calculate noise exposure
Q619. Which material should never be used to absorb mercury spills?
A. Zinc powder
B. Sulfur powder
C. Sawdust
D. Activated charcoal
Q620. Which type of radiation can penetrate deeply into human tissue?

A. Gamma radiation
B. Alpha radiation
C. Beta radiation
D. Ultraviolet radiation

Q621. What is the main goal of an ergonomics program?

A. To adapt the workplace to fit the worker's physical needs
B. To reduce lighting costs
C. To increase work speed
D. To simplify machine design

Q622. Which biological agent is commonly found in poorly maintained HVAC systems?

A. Influenza virus
B. E. coli
C. Salmonella
D. Legionella bacteria

Q623. Which physical hazard is measured in millisieverts (mSv)?

A. Radiation dose
B. Sound pressure
C. Airborne particles
D. Thermal stress

Q624. What is the primary goal of Total Worker Health programs?

A. To integrate safety, wellness, and well-being initiatives
B. To focus only on injury prevention
C. To reduce absenteeism
D. To enforce disciplinary measures

Q625. Which occupational hazard results from exposure to intense visible light?

A. Hand-arm vibration
B. Hearing loss
C. Eye strain and retinal injury
D. Skin sensitization

Q626. Which principle of ISO 14001 ensures continuous improvement in environmental performance?

A. Compliance verification
B. Plan-Do-Check-Act cycle
C. Internal auditing
D. Management review

Q627. What is the main goal of an Environmental Management System (EMS)?

A. Reduce production costs
B. Improve environmental performance
C. Increase employee productivity
D. Enhance product marketing

Q628. Which pollutant is primarily associated with acid rain formation?

A. Sulfur dioxide
B. Methane
C. Carbon dioxide
D. Ozone

Q629. Which process converts organic waste into stable humus-like material?

A. Incineration
B. Composting
C. Filtration
D. Adsorption

Q630. Which gas is a key contributor to the greenhouse effect due to its long atmospheric lifespan?

A. Nitrogen
B. Carbon dioxide
C. Hydrogen
D. Argon

Q631. Which environmental performance indicator measures water reuse efficiency?
A. Water recycling ratio
B. Carbon footprint
C. Energy intensity
D. Air emission index

Q632. Which of the following is the most effective first step in pollution prevention?
A. End-of-pipe treatment
B. Energy recovery
C. Recycling
D. Source reduction

Q633. Which environmental concept focuses on minimizing total lifecycle impacts?
A. Risk management
B. Life Cycle Assessment
C. Quality assurance
D. Energy audit

Q634. Which waste category includes discarded commercial chemical products?
A. F-listed waste
B. K-listed waste
C. P/U-listed waste
D. D-listed waste

Q635. What is the purpose of an environmental impact assessment (EIA)?
A. Identify business opportunities
B. Evaluate potential environmental consequences of projects
C. Promote marketing initiatives
D. Optimize labor efficiency

Q636. Which management principle is emphasized in sustainable resource utilization?
A. Short-term profit focus
B. Waste maximization
C. Resource depletion
D. Conservation and regeneration

Q637. Which ISO standard provides guidelines for environmental management systems?
A. ISO 9001
B. ISO 14001
C. ISO 45001
D. ISO 50001

Q638. Which term refers to sustainable practices across the entire value chain?
A. Green logistics
B. Circular economy
C. End-of-life disposal
D. Linear production

Q639. What is the major environmental concern with persistent organic pollutants (POPs)?
A. Rapid biodegradation
B. High volatility
C. Bioaccumulation and toxicity
D. Lack of reactivity

Q640. Which process is designed to separate particulates from industrial exhaust streams?

A. Distillation
B. Electrostatic precipitation
C. Sedimentation
D. Osmosis

Q641. Which energy source is classified as renewable?
A. Coal
B. Oil
C. Wind
D. Natural gas

Q642. What is the purpose of an environmental audit?
A. Verify environmental performance and compliance
B. Develop employee benefits
C. Improve marketing strategies
D. Enhance product aesthetics

Q643. Which waste management hierarchy is most preferred environmentally?
A. Disposal
B. Recycling
C. Source reduction
D. Treatment

Q644. Which global environmental agreement addresses the depletion of the ozone layer?
A. Kyoto Protocol
B. Montreal Protocol
C. Paris Agreement
D. Basel Convention

Q645. Which sustainability reporting framework emphasizes environmental, social, and governance (ESG) factors?
A. GRI Standards
B. OSHA 1910
C. NFPA 1600
D. ISO 9001

Q646. Which gas is monitored to control indoor air quality and ventilation efficiency?
A. Carbon dioxide
B. Nitrous oxide
C. Methane
D. Ozone

Q647. Which waste management process reduces waste volume through combustion?
A. Composting
B. Incineration
C. Filtration
D. Precipitation

Q648. Which concept emphasizes transparency in corporate environmental disclosures?
A. Greenwashing
B. Diversification
C. Outsourcing
D. ESG reporting

Q649. Which activity best exemplifies environmental sustainability in industrial operations?
A. Continuous equipment operation
B. Waste heat recovery
C. Chemical overuse
D. Resource hoarding

Q650. Which analytical process quantifies emissions from all operational sources?
A. Carbon footprint assessment
B. Material flow analysis
C. Life cycle costing
D. Cost-benefit analysis
Q651. Which environmental concept prioritizes using renewable energy sources to reduce fossil fuel dependence?
A. Resource depletion
B. Energy transition
C. Industrialization
D. Waste minimization
Q652. What is the primary purpose of an Environmental Management Plan (EMP)?
A. Provide general business objectives
B. Outline measures to minimize project environmental impacts
C. Document employee performance
D. Manage corporate finances
Q653. Which environmental practice involves designing buildings to optimize energy and water efficiency?
A. Industrial ergonomics
B. Thermal expansion
C. Green building design
D. Structural engineering
Q654. Which process treats contaminated groundwater by pumping it to the surface for purification?
A. Soil vapor extraction
B. Pump-and-treat
C. Bioremediation
D. Phytoremediation
Q655. What is the main objective of environmental sustainability?
A. Maximize consumption
B. Preserve resources for future generations
C. Prioritize economic growth only
D. Minimize workforce diversity
Q656. Which environmental metric measures the total energy consumed per unit of output?
A. Energy intensity
B. Water footprint
C. Carbon sequestration
D. Biodiversity index
Q657. Which activity supports sustainable waste management in office environments?
A. Double-sided printing
B. Increased paper consumption
C. Uncontrolled storage
D. Frequent disposal
Q658. Which environmental policy tool sets legally binding limits on pollutant emissions?
A. Voluntary reporting
B. Emission standards
C. Environmental labeling
D. Community outreach
Q659. Which international agreement aims to stabilize greenhouse gas concentrations?
A. Rotterdam Convention
B. Basel Convention

C. Paris Agreement
D. Kyoto Protocol
Q660. Which waste disposal method is most effective for minimizing landfill dependency?
A. Open dumping
B. Waste-to-energy conversion
C. Deep-well injection
D. Landfilling
Q661. Which environmental monitoring parameter helps track acidification in aquatic systems?
A. Dissolved oxygen
B. pH level
C. Conductivity
D. Turbidity
Q662. What is the primary purpose of ISO 14001 certification?
A. Demonstrate financial compliance
B. Enhance personal safety
C. Increase market monopoly
D. Validate environmental management system effectiveness
Q663. Which method prevents soil erosion most effectively on steep slopes?
A. Clear cutting
B. Contour plowing
C. Overgrazing
D. Landfilling
Q664. Which environmental practice encourages circular material use and minimal waste?
A. Linear consumption
B. Cradle-to-cradle design
C. Disposal planning
D. Excess packaging
Q665. Which environmental strategy focuses on adapting to climate change effects?
A. Mitigation
B. Adaptation
C. Remediation
D. Conservation
Q666. Which pollutant is most responsible for smog formation?
A. Nitrogen oxides
B. Lead
C. Carbon monoxide
D. Hydrogen sulfide
Q667. Which energy efficiency measure reduces building cooling loads?
A. Dark roof coatings
B. Reflective roofing
C. Excess lighting
D. Increased equipment heat
Q668. Which air pollutant primarily results from vehicle exhaust?
A. Carbon monoxide
B. Radon
C. Lead dust
D. Sulfur dioxide
Q669. What is the environmental purpose of a spill prevention plan?
A. Reduce spill response time
B. Increase chemical use

C. Prevent and control accidental releases
D. Enhance storage capacity
Q670. Which resource conservation practice involves using less water-intensive technologies?
A. Water efficiency
B. Carbon sequestration
C. Energy diversification
D. Land reclamation
Q671. Which document establishes environmental objectives and commitments within an organization?
A. Environmental policy
B. Incident log
C. Training manual
D. Purchasing order
Q672. Which pollutant is most commonly associated with acid rain and respiratory issues?
A. Oxygen
B. Ammonia
C. Methane
D. Sulfur dioxide
Q673. Which waste classification covers ignitable, corrosive, reactive, or toxic materials?
A. Hazardous waste
B. Municipal waste
C. Construction debris
D. Inert solids
Q674. Which energy management system standard focuses on energy performance improvement?
A. ISO 9001
B. ISO 45001
C. ISO 50001
D. ISO 14001
Q675. Which analysis evaluates environmental risks associated with industrial processes?
A. Failure mode analysis
B. Environmental risk assessment
C. Market feasibility study
D. Productivity analysis
Q676. Which principle of adult learning emphasizes learners' need to understand the relevance of training?
A. Feedback
B. Motivation
C. Readiness to learn
D. Orientation to learning
Q677. Which model is most commonly used to structure the training development process?
A. ADDIE model
B. PDSA model
C. DMAIC model
D. RACI model
Q678. According to Bloom's Taxonomy, which cognitive level involves applying learned material in new situations?
A. Understanding
B. Application
C. Analysis
D. Evaluation

Q679. Which delivery method is most effective for hands-on skill development?
A. Lecture
B. Simulation
C. Webinar
D. Reading assignment
Q680. What is the primary purpose of a training needs assessment?
A. Assign trainers
B. Determine budgets
C. Identify performance gaps
D. Develop schedules
Q681. Which principle of learning emphasizes participation and engagement?
A. Feedback
B. Active learning
C. Reinforcement
D. Reflection
Q682. Which Kirkpatrick evaluation level measures changes in behavior after training?
A. Level 1 – Reaction
B. Level 2 – Learning
C. Level 3 – Behavior
D. Level 4 – Results
Q683. Which instructional method provides immediate feedback through interaction?
A. Self-paced reading
B. Computer-based training
C. Peer mentoring
D. Lecture
Q684. What is the most important factor in ensuring the success of safety training?
A. Technical complexity
B. Learner engagement
C. Course duration
D. Facility size
Q685. Which element is most critical in the design phase of the ADDIE model?
A. Evaluation criteria
B. Trainer selection
C. Schedule planning
D. Documentation
Q686. Which adult learning characteristic emphasizes prior experience as a learning resource?
A. Experience-based orientation
B. Motivation to learn
C. Learning readiness
D. Goal orientation
Q687. Which instructional tool enhances visual learning and data comprehension?
A. Whiteboard notes
B. Charts and infographics
C. Verbal repetition
D. Written manuals
Q688. What is the main goal of safety communication programs?
A. Regulatory compliance
B. Awareness and behavior change
C. Branding
D. Documentation control

Q689. Which learning domain focuses on skill development through practice?
A. Affective
B. Cognitive
C. Psychomotor
D. Behavioral
Q690. Which feedback type most effectively reinforces positive learner performance?
A. Delayed feedback
B. Constructive feedback
C. Punitive feedback
D. Minimal feedback
Q691. Which communication method enhances retention by combining visual and auditory input?
A. Reading assignments
B. Role-play
C. Group discussion
D. Demonstration
Q692. What is the purpose of conducting post-training evaluations?
A. Measure long-term retention and performance impact
B. Identify new training topics
C. Replace needs analysis
D. Assess trainer attendance
Q693. Which factor most influences adult learner motivation?
A. Peer competition
B. Practical relevance
C. External rewards
D. Mandatory attendance
Q694. Which step of the communication process ensures message understanding?
A. Encoding
B. Decoding
C. Feedback
D. Transmission
Q695. Which evaluation level in Kirkpatrick's model assesses organizational benefits?
A. Level 1 – Reaction
B. Level 2 – Learning
C. Level 3 – Behavior
D. Level 4 – Results
Q696. Which approach is most effective for training employees on emergency procedures?
A. Reading policy documents
B. Classroom lectures
C. Practical drills and exercises
D. Online modules
Q697. Which training strategy enhances long-term knowledge retention?
A. Cram sessions
B. Spaced learning
C. One-time lectures
D. Memorization
Q698. Which factor best determines the effectiveness of a communication channel?
A. Message volume
B. Speed of transmission
C. Audience understanding
D. Number of participants

Q699. Which instructional design model emphasizes continuous feedback and flexibility?
A. ADDIE
B. SAM (Successive Approximation Model)
C. Kirkpatrick
D. Bloom's Taxonomy
Q700. Which communication method encourages open dialogue between employees and management?
A. Anonymous surveys
B. Two-way communication
C. Written memos
D. Annual reports

Answer keys and Explanations (601-700)

Q601. Answer: B

Explanation:

Effective ventilation and dust suppression systems prevent inhalation of harmful particles that cause respiratory diseases like pneumoconiosis. These systems control airborne dust at the source, ensuring cleaner air. When combined with regular monitoring and appropriate respiratory protection, they significantly reduce long-term exposure risks and improve overall workplace air quality.

Q602. Answer: A

Explanation:

Prolonged use of vibrating hand tools transmits mechanical energy to the hands and arms, leading to nerve and blood vessel damage. This condition, known as hand-arm vibration syndrome, causes numbness, tingling, and reduced grip strength. Using anti-vibration tools, gloves, and limiting exposure time effectively prevents the onset of this occupational disease.

Q603. Answer: A

Explanation:

The inverse square law states that radiation intensity decreases proportionally to the square of the distance from the source. Doubling the distance reduces exposure by a factor of four. Maintaining distance is a simple and highly effective method of minimizing exposure to ionizing radiation in workplaces where such hazards exist.

Q604. Answer: C

Explanation:

Local exhaust ventilation captures contaminants such as fumes, dust, or vapors directly at their point of generation before they disperse into the air. This targeted control prevents worker exposure, improves indoor air quality, and enhances overall occupational health and safety in industrial environments.

Q605. Answer: A

Explanation:

Chemical absorption through the skin depends on the integrity of the skin and the solubility of the substance. Damaged or moist skin allows faster penetration, while lipid-soluble chemicals are absorbed more easily. Maintaining skin health and using protective gloves prevent harmful absorption and systemic toxicity.

Q606. Answer: D

Explanation:

Noise-induced hearing loss results from prolonged exposure to high sound levels that damage the inner ear's sensory cells. The condition is irreversible but preventable through proper engineering controls, hearing protection, and routine audiometric testing. Early intervention and consistent noise monitoring ensure long-term hearing preservation for workers.

Q607. Answer: A

Explanation:

Integrated sampling collects airborne contaminants over several hours to determine average exposure levels during a shift. This approach provides representative data for chronic exposure evaluation, ensuring compliance with occupational exposure limits and guiding effective control strategies for sustained worker protection.

Q608. Answer: B

Explanation:

An occupational health nurse supports workplace health by conducting medical surveillance, promoting wellness, managing injuries, and educating employees on safe work practices. Their preventive focus helps detect early signs of illness and ensures a healthier, more productive workforce.

Q609. Answer: A

Explanation:

The Time-Weighted Average (TWA) reflects the average noise exposure over a typical eight-hour shift, accounting for varying sound levels. It determines compliance with noise exposure limits and guides the need for hearing conservation programs and engineering noise control interventions.

Q610. Answer: A

Explanation:

Active air sampling uses pumps to draw air through sorbent tubes that trap airborne contaminants for laboratory analysis. This technique accurately quantifies exposure concentrations, ensuring occupational exposure remains below permissible limits and guiding effective control measures.

Q611. Answer: C

Explanation:

Heat stroke occurs when the body's cooling mechanisms fail, causing a dangerous rise in core temperature. Symptoms include confusion, hot skin, and potential organ failure. Immediate cooling, hydration, and medical attention are critical. Preventing heat stroke involves hydration, acclimatization, and environmental monitoring in high-temperature workplaces.

Q612. Answer: A

Explanation:

A respirator protects workers by filtering or supplying breathable air when airborne contaminants are present. Proper selection, fit testing, and maintenance ensure it functions effectively. Respiratory protection is vital when engineering controls cannot sufficiently reduce exposure levels.

Q613. Answer: A

Explanation:

Carbon monoxide binds with hemoglobin more readily than oxygen, preventing oxygen transport in the bloodstream. This leads to tissue hypoxia, dizziness, and potentially fatal outcomes. Adequate ventilation and continuous monitoring are essential to prevent carbon monoxide poisoning in workplaces.

Q614. Answer: D

Explanation:

Epoxy resins can cause skin sensitization, resulting in allergic reactions after repeated contact. Once sensitized, even minor exposure can trigger severe dermatitis. Using protective gloves, barrier creams, and proper hygiene prevents sensitization and ensures long-term skin health.

Q615. Answer: A

Explanation:

A hazard communication program ensures that employees understand the risks associated with chemicals they handle through labeling, Safety Data Sheets, and training. It empowers workers to follow safe practices and respond appropriately to potential chemical hazards.

Q616. Answer: B

Explanation:

Continuous high-intensity noise causes irreversible damage to sensory cells within the inner ear. This form of exposure is more dangerous than short-term or low-level noise. Implementing engineering controls and using hearing protection minimize the risk of noise-induced hearing loss.

Q617. Answer: A

Explanation:

Inhalation absorption occurs when airborne toxins enter the lungs and pass into the bloodstream, quickly affecting vital organs. Effective ventilation, respiratory protection, and exposure monitoring prevent excessive inhalation and ensure workers remain within safe limits.

Q618. Answer: A

Explanation:

A Biological Exposure Index (BEI) quantifies the concentration of a substance or its metabolites in biological samples such as blood or urine. It helps determine whether workplace controls are adequate and ensures exposure remains within safe biological limits.

Q619. Answer: C

Explanation:

Sawdust reacts with mercury, increasing vaporization and worsening contamination. Special mercury spill kits or sulfur powder should be used to bind and contain mercury safely. Proper cleanup prevents toxic vapor exposure and protects workers from poisoning.

Q620. Answer: A

Explanation:

Gamma radiation is highly penetrating, capable of passing through human tissue and many materials. Shielding with dense substances like lead and maintaining safe distance are vital for minimizing exposure and protecting workers from harmful ionizing radiation effects.

Q621. Answer: A

Explanation:

Ergonomics focuses on designing work tasks and equipment to match human capabilities and limitations. Proper ergonomic practices prevent musculoskeletal injuries, improve comfort, and enhance productivity by reducing strain, repetitive motion, and awkward postures.

Q622. Answer: D

Explanation:

Legionella bacteria thrive in warm, stagnant water within HVAC cooling systems and can cause Legionnaires' disease when inhaled. Regular maintenance, disinfection, and temperature control prevent bacterial growth and ensure healthy indoor air quality.

Q623. Answer: A

Explanation:

The millisievert (mSv) is a unit used to measure the absorbed dose of ionizing radiation in biological tissue. Monitoring radiation dose ensures compliance with exposure limits and safeguards workers' health by preventing radiation-related diseases.

Q624. Answer: A

Explanation:

Total Worker Health programs combine occupational safety with health promotion efforts to improve overall worker well-being. They address lifestyle, environmental, and psychosocial factors, creating a holistic approach that enhances productivity and long-term wellness.

Q625. Answer: C

Explanation:

Exposure to intense visible light, such as from welding arcs or lasers, causes eye strain, temporary blindness, or retinal burns. Proper shielding, eye protection, and work practice controls protect workers from these serious visual hazards.

Q626. Answer: B

Explanation:

The Plan-Do-Check-Act cycle is the foundation of ISO 14001's structure, driving continuous environmental performance improvement. It integrates planning, implementation, evaluation, and corrective action to ensure an organization continually refines its environmental management system.

This cycle ensures environmental objectives are met effectively through systematic monitoring and responsive action.

Q627. Answer: B

Explanation:

An Environmental Management System aims to enhance environmental performance by managing impacts systematically. It helps organizations identify, control, and reduce environmental risks while promoting sustainability and regulatory compliance. By structuring continuous evaluation and improvement, the EMS supports long-term environmental and operational excellence.

Q628. Answer: A

Explanation:

Sulfur dioxide, released from burning fossil fuels, reacts in the atmosphere with water vapor and oxygen to form sulfuric acid. This process leads to acid rain, which damages ecosystems, structures, and soil quality. Controlling sulfur emissions is crucial for minimizing acid rain's environmental effects.

Q629. Answer: B

Explanation:

Composting decomposes organic matter through microbial activity, transforming waste into nutrient-rich humus. This natural process stabilizes waste, reduces landfill volume, and supports soil health. Proper temperature and oxygen control ensure efficient decomposition and pathogen reduction, producing a valuable soil conditioner.

Q630. Answer: B

Explanation:

Carbon dioxide is a major greenhouse gas that traps heat in the Earth's atmosphere, leading to climate change. Its long lifespan allows it to accumulate over decades, amplifying global warming. Reducing CO_2 emissions through cleaner energy sources and carbon management is essential for climate mitigation.

Q631. Answer: A

Explanation:

The water recycling ratio quantifies the proportion of total water that is reused within operations, indicating conservation efficiency. A higher ratio reflects effective water stewardship, minimizing withdrawals from natural sources. This metric supports sustainability objectives and regulatory compliance in industries with significant water consumption.

Q632. Answer: D

Explanation:

Source reduction prevents waste and pollutants from being created at the outset. It minimizes environmental impact by redesigning processes, substituting materials, or optimizing resource use. This proactive approach aligns with sustainable manufacturing principles, reducing treatment costs and compliance burdens downstream.

Q633. Answer: B

Explanation:

Life Cycle Assessment evaluates environmental impacts of a product from raw material extraction to disposal. It provides insights into resource use, emissions, and energy consumption, identifying opportunities for sustainability improvement. LCA helps organizations make informed decisions that reduce ecological footprints.

Q634. Answer: C

Explanation:

P- and U-listed wastes consist of unused commercial chemical products that are hazardous when discarded. They pose environmental and health risks if not managed properly. These wastes require special labeling, handling, and disposal procedures to ensure regulatory compliance and prevent contamination.

Q635. Answer: B

Explanation:

An Environmental Impact Assessment systematically evaluates potential effects of proposed projects on the environment. It informs decision-makers by identifying mitigation measures to prevent harm to air, water, land, and ecosystems. EIA supports sustainable development and regulatory compliance.

Q636. Answer: D

Explanation:

Sustainable resource utilization focuses on conservation and regeneration, ensuring resources are used efficiently and replenished for future generations. This principle balances economic needs with ecological protection, promoting responsible production and consumption practices that maintain ecosystem integrity.

Q637. Answer: B

Explanation:

ISO 14001 defines requirements for establishing, implementing, maintaining, and improving an Environmental Management System. It helps organizations achieve environmental objectives through systematic control of processes, compliance, and continual improvement. This standard supports sustainable operations and global environmental accountability.

Q638. Answer: B

Explanation:

The circular economy promotes closed-loop systems where materials are reused, recycled, or repurposed to minimize waste. It replaces linear consumption models, enhancing resource efficiency and environmental protection. This approach supports long-term sustainability and reduces dependency on virgin resources.

Q639. Answer: C

Explanation:

Persistent organic pollutants resist degradation and accumulate in living organisms, causing toxic effects through the food chain. Their long-term persistence and global transport potential make them significant ecological and health threats, requiring strict control and international regulation.

Q640. Answer: B

Explanation:

Electrostatic precipitators remove fine particulates from exhaust gases using electrically charged plates. The process captures dust and smoke efficiently, reducing air pollution. This technology is vital in power plants and manufacturing facilities for emission control and regulatory compliance.

Q641. Answer: C

Explanation:

Wind energy is renewable because it harnesses natural atmospheric motion without depleting resources. It produces electricity with zero direct emissions and contributes to sustainable energy transition. Wind power supports decarbonization and energy diversification goals globally.

Q642. Answer: A

Explanation:

An environmental audit evaluates an organization's compliance with environmental regulations and internal policies. It identifies improvement areas, ensures operational control, and confirms effectiveness of the Environmental Management System. Regular audits strengthen environmental accountability and risk management.

Q643. Answer: C

Explanation:

Source reduction prevents waste generation by minimizing material use and redesigning processes. It is prioritized over recycling or disposal as it eliminates pollution at its origin. This strategy supports resource conservation and long-term sustainability.

Q644. Answer: B

Explanation:

The Montreal Protocol regulates substances that deplete the ozone layer, such as chlorofluorocarbons. It successfully phased out many harmful compounds, contributing to ozone recovery. This treaty remains one of the most effective international environmental agreements.

Q645. Answer: A

Explanation:

The Global Reporting Initiative (GRI) Standards provide a comprehensive framework for reporting ESG performance. They guide organizations in disclosing environmental, social, and governance impacts transparently, promoting stakeholder trust and accountability. GRI supports responsible decision-making and sustainability integration.

Q646. Answer: A

Explanation:

Carbon dioxide levels indicate ventilation effectiveness and occupancy in indoor environments. Elevated concentrations suggest inadequate airflow and potential discomfort. Monitoring CO_2 helps maintain healthy indoor air quality and supports energy-efficient ventilation management.

Q647. Answer: B

Explanation:

Incineration burns solid waste at high temperatures, reducing its volume and generating energy. It provides effective disposal for non-recyclable materials, minimizing landfill use. Proper emission controls ensure compliance and minimize air pollution impacts.

Q648. Answer: D

Explanation:

ESG reporting ensures organizations communicate their environmental, social, and governance impacts clearly. It promotes accountability, stakeholder engagement, and sustainable decision-making. Transparent ESG disclosure enhances investor confidence and aligns business practices with long-term environmental responsibility.

Q649. Answer: B

Explanation:

Waste heat recovery captures and reuses thermal energy from processes, reducing fuel consumption and emissions. It increases energy efficiency and operational sustainability. This practice aligns with green manufacturing goals and resource optimization.

Q650. Answer: A

Explanation:

A carbon footprint assessment quantifies greenhouse gas emissions from an organization's activities, products, or services. It identifies emission hotspots and guides reduction strategies. Measuring carbon output is critical for achieving sustainability goals and complying with environmental standards.

Q651. Answer: B

Explanation:

Energy transition focuses on shifting from fossil fuels to renewable sources such as solar, wind, and hydro power. This shift reduces greenhouse gas emissions, supports sustainability goals, and ensures long-term energy security. Transitioning to renewables is a cornerstone of climate mitigation and environmental management efforts worldwide.

Q652. Answer: B

Explanation:

An Environmental Management Plan details strategies to mitigate adverse environmental impacts during project implementation. It specifies monitoring, control, and mitigation measures for air, water, and waste. EMPs ensure compliance with regulations and promote sustainable project execution through structured environmental safeguards.

Q653. Answer: C

Explanation:

Green building design integrates sustainable materials, efficient lighting, ventilation, and water systems to minimize environmental impact. It enhances occupant comfort, reduces resource use, and supports long-term operational savings. Such designs are vital for reducing carbon footprints and supporting eco-friendly infrastructure.

Q654. Answer: B

Explanation:

The pump-and-treat method removes contaminated groundwater, treats it above ground, and returns clean water to the aquifer or discharge system. This process effectively reduces pollutant concentrations and restores groundwater quality. It is a key remediation technique for sites impacted by hazardous chemicals.

Q655. Answer: B

Explanation:

Environmental sustainability ensures natural resources are conserved and managed responsibly to meet current needs without compromising future generations. It promotes balanced development by integrating environmental, social, and economic goals, fostering long-term ecological resilience.

Q656. Answer: A

Explanation:

Energy intensity evaluates how efficiently an organization converts energy into output. A lower intensity indicates higher efficiency and reduced environmental impact. Monitoring this metric supports sustainability goals by identifying opportunities to minimize energy waste and operational costs.

Q657. Answer: A

Explanation:

Double-sided printing reduces paper use and waste generation, supporting sustainability and cost reduction. This simple practice aligns with waste minimization strategies by conserving resources and lowering landfill contributions. Implementing similar conservation habits builds environmental awareness across the workplace.

Q658. Answer: B

Explanation:

Emission standards establish enforceable limits on pollutants released into air, water, or soil. These regulations ensure compliance, protect public health, and drive technological innovation. Meeting emission standards reflects effective environmental management and operational responsibility.

Q659. Answer: C

Explanation:

The Paris Agreement seeks to limit global temperature rise by reducing greenhouse gas emissions worldwide. Participating countries commit to nationally determined contributions and transparent progress reporting. It represents a unified global effort to mitigate climate change and promote resilience.

Q660. Answer: B

Explanation:

Waste-to-energy conversion incinerates solid waste to produce electricity or heat, reducing landfill use. It transforms waste into a renewable energy source, conserving space and minimizing methane emissions. Properly managed facilities ensure emission control and energy recovery efficiency.

Q661. Answer: B

Explanation:

pH level monitoring detects acidification in aquatic systems caused by acid rain or industrial discharges. Maintaining balanced pH is crucial for aquatic organism survival and ecosystem stability. Regular monitoring supports early detection and pollution control efforts.

Q662. Answer: D

Explanation:

ISO 14001 certification confirms an organization's environmental management system meets international standards. It demonstrates commitment to environmental responsibility, continual improvement, and regulatory compliance. Certification enhances corporate reputation and operational performance.

Q663. Answer: B

Explanation:

Contour plowing follows natural land contours to reduce water runoff and soil erosion. It improves water retention and promotes sustainable agriculture. This conservation technique protects soil fertility and prevents sediment pollution in nearby waterways.

Q664. Answer: B

Explanation:

Cradle-to-cradle design ensures materials are continuously reused or recycled, creating closed production loops. It minimizes waste and resource depletion, promoting long-term sustainability. This design philosophy supports innovation and responsible material management.

Q665. Answer: B

Explanation:

Adaptation strategies enhance resilience by preparing for and adjusting to climate impacts such as floods, droughts, and temperature changes. They include infrastructure upgrades, early warning systems, and policy planning. Adaptation complements mitigation efforts for comprehensive climate response.

Q666. Answer: A

Explanation:

Nitrogen oxides combine with volatile organic compounds under sunlight to form photochemical smog. Smog contributes to respiratory issues and reduced visibility. Controlling NOx emissions from vehicles and industries is crucial for improving urban air quality.

Q667. Answer: B

Explanation:

Reflective roofing materials reflect sunlight and absorb less heat, lowering indoor temperatures and reducing cooling energy demand. This sustainable design feature enhances occupant comfort and decreases air conditioning use, supporting energy efficiency goals.

Q668. Answer: A

Explanation:

Carbon monoxide is produced by incomplete fuel combustion in vehicle engines. It poses health hazards and contributes to air pollution. Emission control technologies like catalytic converters help reduce CO output and improve urban air quality.

Q669. Answer: C

Explanation:

A spill prevention plan outlines procedures to avoid accidental chemical releases and minimize impact if they occur. It includes containment, inspection, and emergency response measures. Effective planning protects the environment and ensures compliance with regulations.

Q670. Answer: A

Explanation:

Water efficiency minimizes consumption by adopting technologies and processes that achieve the same outcomes using less water. It supports sustainability by conserving freshwater resources and reducing energy used for treatment and distribution.

Q671. Answer: A

Explanation:

An environmental policy formally states an organization's commitment to managing environmental impacts and achieving sustainability goals. It provides a foundation for objectives, procedures, and continual improvement. The policy guides actions and promotes accountability.

Q672. Answer: D

Explanation:

Sulfur dioxide emissions combine with atmospheric moisture to form sulfuric acid, leading to acid rain and respiratory irritation. Reducing SO_2 emissions through cleaner fuels and scrubbing systems mitigates these environmental and health impacts.

Q673. Answer: A

Explanation:

Hazardous waste includes substances that pose potential harm due to ignitability, corrosivity, reactivity, or toxicity. These materials require special handling and disposal to prevent pollution and ensure regulatory compliance under environmental laws.

Q674. Answer: C

Explanation:

ISO 50001 provides a framework for systematic energy management to enhance performance and efficiency. It guides organizations in monitoring, reducing, and optimizing energy use while supporting cost savings and sustainability targets.

Q675. Answer: B

Explanation:

Environmental risk assessment identifies, evaluates, and prioritizes potential environmental hazards from operations or projects. It supports proactive control measures and informed decision-making. Conducting this analysis ensures environmental protection and regulatory compliance.

Q676. Answer: D

Explanation:

Orientation to learning stresses that adults are problem-centered and need to see the immediate value and application of new knowledge. This principle ensures that training content connects directly to real work tasks, enhancing motivation and engagement. It makes learning purposeful and outcome-focused, improving retention and performance in the workplace.

Q677. Answer: A

Explanation:

The ADDIE model—Analysis, Design, Development, Implementation, and Evaluation—provides a systematic framework for creating effective training. It ensures each stage builds upon the previous one to meet learning objectives. By integrating evaluation at every phase, ADDIE promotes continuous improvement and alignment with organizational needs.

Q678. Answer: B

Explanation:

The application level in Bloom's Taxonomy focuses on using acquired knowledge or skills in new or practical contexts. It measures a learner's ability to transfer learning beyond theory into action. This level ensures training produces observable performance improvements and supports workplace competency development.

Q679. Answer: B

Explanation:

Simulation training replicates real-world scenarios, allowing learners to practice skills in a controlled environment. It enhances experiential learning and builds confidence through repetition without real risk. This method is especially effective for safety, equipment operation, and emergency response training programs.

Q680. Answer: C

Explanation:

A training needs assessment identifies gaps between current and desired performance. It helps organizations determine which skills, knowledge, or behaviors require development to achieve goals. Conducting this analysis ensures resources target critical areas that directly influence productivity and safety outcomes.

Q681. Answer: B

Explanation:

Active learning involves learner participation through discussions, problem-solving, and exercises. It promotes understanding, retention, and practical skill application by engaging learners cognitively and physically. This approach transforms passive instruction into interactive experiences that enhance comprehension and motivation.

Q682. Answer: C

Explanation:

Kirkpatrick's Level 3 evaluation assesses whether participants apply what they learned to their job roles. It measures behavioral change and skill transfer resulting from training. Monitoring this level demonstrates training effectiveness in improving performance and influencing workplace outcomes.

Q683. Answer: B

Explanation:

Computer-based training allows learners to engage interactively, receiving instant feedback on responses and performance. This self-directed format supports personalized learning and reinforces correct actions. The feedback mechanism enhances retention and ensures comprehension before progression to new topics.

Q684. Answer: B

Explanation:

Learner engagement ensures that participants actively process and apply safety concepts. Engaged learners are more likely to retain information and modify behaviors that improve workplace safety. Interactive and relevant instruction drives motivation, accountability, and successful learning outcomes.

Q685. Answer: A

Explanation:

During the design phase, establishing evaluation criteria ensures learning outcomes align with organizational goals and measurable performance standards. It provides the foundation for assessing training success and helps structure content, materials, and methods effectively to achieve desired competencies.

Q686. Answer: A

Explanation:

Experience-based orientation recognizes that adult learners bring valuable prior knowledge that influences how they acquire new skills. Trainers should connect new information to existing experiences, facilitating meaningful understanding and application. This approach validates learner expertise and enhances engagement.

Q687. Answer: B

Explanation:

Charts and infographics visually simplify complex data, making it easier for learners to interpret and retain key concepts. Visual aids strengthen comprehension by linking abstract information to clear imagery. They enhance engagement and support diverse learning preferences in group settings.

Q688. Answer: B

Explanation:

Safety communication aims to raise awareness and influence behavior that reduces risk. Clear, consistent messages help employees recognize hazards and act safely. Effective communication builds a proactive safety culture that fosters responsibility and participation across all levels of the organization.

Q689. Answer: C

Explanation:

The psychomotor domain relates to physical skill development through observation, imitation, and practice. It measures a learner's ability to perform specific tasks accurately and efficiently. This domain is essential for training requiring hands-on proficiency, such as equipment handling and emergency response.

Q690. Answer: B

Explanation:

Constructive feedback highlights strengths while providing guidance for improvement in a supportive manner. It builds confidence and encourages continuous learning. This approach maintains motivation and ensures learners remain engaged and receptive to development opportunities.

Q691. Answer: D

Explanation:

Demonstrations engage both sight and hearing, reinforcing understanding through multiple sensory channels. Observing a task performed correctly enables learners to replicate it accurately. This multisensory approach supports faster skill acquisition and higher retention rates.

Q692. Answer: A

Explanation:

Post-training evaluations assess how well learners retain and apply skills after training. They measure real-world effectiveness, identifying areas for reinforcement or improvement. Ongoing evaluation supports continuous program enhancement and ensures sustained behavior change.

Q693. Answer: B

Explanation:

Adults are motivated when learning addresses real-world challenges or immediate job needs. Practical relevance ensures they see clear benefits and applicability of training. Connecting content to daily tasks enhances motivation and promotes long-term retention.

Q694. Answer: C

Explanation:

Feedback confirms that the receiver understood the sender's message correctly. It closes the communication loop, enabling clarification and improvement. Effective feedback reduces miscommunication and promotes clear, two-way information exchange in workplace training environments.

Q695. Answer: D

Explanation:

Level 4 of Kirkpatrick's model measures the broader results of training, such as productivity gains, reduced incidents, or cost savings. It links learning outcomes to organizational performance improvements, demonstrating training's overall value and impact.

Q696. Answer: C

Explanation:

Practical drills simulate real emergencies, allowing employees to apply procedures in realistic scenarios. This experiential learning enhances retention, confidence, and quick decision-making under pressure. Drills ensure readiness and strengthen response capability in actual emergencies.

Q697. Answer: B

Explanation:

Spaced learning reinforces information over multiple sessions, allowing time for review and consolidation. This method strengthens memory retention and reduces knowledge decay compared to single-session instruction. It supports sustainable learning outcomes and performance improvement.

Q698. Answer: C

Explanation:

A communication channel is effective when the audience comprehends the intended message accurately. Clarity, simplicity, and feedback mechanisms enhance understanding and reduce distortion. Selecting the right channel ensures alignment between sender intent and receiver interpretation.

Q699. Answer: B

Explanation:

The Successive Approximation Model (SAM) promotes iterative design with continuous testing and feedback loops. It allows adjustments during development, creating dynamic and learner-centered training. SAM ensures adaptability and responsiveness to changing learner needs or feedback.

Q700. Answer: B

Explanation:

Two-way communication fosters trust by allowing feedback, questions, and dialogue between employees and leaders. It ensures transparency, strengthens relationships, and enhances organizational engagement. This open exchange supports continuous improvement and collaborative problem-solving.

Questions (701-800)

Q701. Which factor most enhances message clarity in workplace communication?
A. Technical jargon
B. Concise language
C. Complex terminology
D. Ambiguous phrasing

Q702. Which instructional technique enhances learning by connecting new material to existing knowledge?
A. Reinforcement
B. Scaffolding
C. Simulation
D. Assessment

Q703. Which method best develops communication skills for supervisors?
A. Simulation role-play
B. Technical manuals
C. Online quizzes
D. Classroom lectures

Q704. Which training evaluation technique directly measures learner performance on the job?
A. Written test
B. Observation
C. Survey
D. Self-assessment

Q705. What is the key advantage of blended learning?
A. Reduced learner accountability
B. Elimination of instructor feedback
C. Increased complexity
D. Combination of online and in-person methods

Q706. Which factor contributes most to effective team communication during safety meetings?
A. Hierarchical control
B. Open participation
C. Rigid agendas
D. Monologues

Q707. Which adult learning theory highlights self-directed learning and internal motivation?
A. Behaviorism
B. Andragogy
C. Constructivism
D. Pedagogy

Q708. Which feedback method is most effective for skill-based training?
A. Immediate corrective feedback
B. Delayed group summary
C. Written test review
D. Periodic surveys

Q709. Which component of an effective training program defines measurable learning outcomes?
A. Course title
B. Learning objectives
C. Instructor biography
D. Session schedule

Q710. Which form of reinforcement strengthens positive behavior in the workplace?
A. Ignoring performance

B. Public criticism
C. Constructive praise
D. Task repetition

Q711. What is the main advantage of microlearning in training programs?
A. In-depth analysis
B. Short, focused learning segments
C. Extended reading sessions
D. Group lectures

Q712. Which principle ensures information is retained by linking learning to real experiences?
A. Abstraction
B. Relevance
C. Observation
D. Repetition

Q713. Which communication barrier most often disrupts message accuracy?
A. Active listening
B. Eye contact
C. Feedback
D. Noise

Q714. Which training delivery method best supports geographically dispersed teams?
A. E-learning
B. In-person workshop
C. Printed manuals
D. On-site sessions

Q715. Which concept describes employees' belief in their ability to perform tasks successfully?
A. Self-efficacy
B. Motivation
C. Competence
D. Awareness

Q716. Which communication approach best supports a strong safety culture?
A. One-way announcements
B. Open dialogue and feedback
C. Restrictive reporting
D. Technical bulletins only

Q717. Which element of nonverbal communication conveys confidence and attentiveness?
A. Crossed arms
B. Eye contact
C. Turning away
D. Fidgeting

Q718. Which evaluation method determines whether training improves job performance metrics?
A. Pre-test only
B. Knowledge quiz
C. Return on investment analysis
D. Satisfaction survey

Q719. Which strategy helps maintain learning effectiveness in virtual training environments?
A. Passive listening
B. Interactive engagement tools
C. Long monologues
D. Text-heavy slides

Q720. Which adult learning characteristic supports problem-solving orientation?
A. Task focus

B. Goal direction
C. Motivation
D. Experience sharing
Q721. What is the primary role of a facilitator in adult learning?
A. Deliver lectures
B. Direct information flow
C. Guide learning experiences
D. Evaluate attendance
Q722. Which teaching method encourages reflective learning and self-assessment?
A. Drill and practice
B. Journaling
C. Lecture
D. Demonstration
Q723. Which element ensures safety messages are culturally inclusive?
A. Technical precision
B. Standardized tone
C. Symbolic complexity
D. Language adaptation
Q724. Which communication style best motivates employees toward safety compliance?
A. Authoritative
B. Collaborative
C. Indifferent
D. Reactive
Q725. Which learning approach emphasizes adapting training to learner needs and progress?
A. Personalized learning
B. Static instruction
C. Group lecturing
D. One-way communication
Q726. Which technique improves retention by varying learning activities?
A. Monotony
B. Interleaving
C. Memorization
D. Rote repetition
Q727. Which communication element is essential for ensuring message understanding across all organizational levels?
A. Feedback loop
B. Encoding
C. Noise reduction
D. Channel selection
Q728. Which adult learning theory emphasizes learning through observation and imitation?
A. Cognitive theory
B. Behavioral theory
C. Social learning theory
D. Constructivist theory
Q729. Which evaluation level measures learners' satisfaction with the training experience?
A. Level 1 – Reaction
B. Level 2 – Learning
C. Level 3 – Behavior
D. Level 4 – Results
Q730. Which communication strategy builds trust and transparency during organizational change?

A. Limited disclosure
B. Frequent, honest updates
C. One-time announcements
D. Avoiding employee input

Q731. Which learning environment best supports active collaboration and critical thinking?
A. Lecture-based
B. Recorded videos
C. Self-paced reading
D. Interactive workshops

Q732. Which communication tool is best for tracking employee training progress?
A. Bulletin board
B. Learning management system
C. Paper checklist
D. Verbal reminders

Q733. Which concept emphasizes adapting training materials to multiple learning styles?
A. Standardization
B. Customization
C. Uniform instruction
D. Simplification

Q734. Which communication format provides the most immediate response?
A. Email
B. Phone call
C. Formal report
D. Memo

Q735. Which training principle ensures alignment between objectives and content delivery?
A. Relevance
B. Consistency
C. Validity
D. Alignment

Q736. Which communication strategy enhances understanding during multicultural team meetings?
A. Fast-paced delivery
B. Using idioms
C. Simplifying language
D. Complex terminology

Q737. Which learning strategy involves learners discovering knowledge through exploration?
A. Guided discovery
B. Lecture-based instruction
C. Passive observation
D. Demonstration

Q738. Which communication skill is most essential for effective listening?
A. Interrupting to clarify
B. Active listening
C. Selective attention
D. Rapid response

Q739. Which factor most influences the success of cross-departmental training?
A. Collaboration between departments
B. Isolated learning
C. Competitive focus
D. Rigid scheduling

Q740. Which delivery format offers flexibility and supports self-paced learning?

A. E-learning
B. Seminar
C. Workshop
D. Field training

Q741. Which nonverbal signal often indicates engagement during conversation?
A. Silence
B. Looking away
C. Crossing arms
D. Nodding

Q742. Which type of training focuses on improving leadership communication and influence?
A. Technical training
B. Soft skills training
C. Compliance training
D. Task-specific training

Q743. Which principle ensures adult learners can control their learning process?
A. Structure
B. Autonomy
C. Authority
D. Evaluation

Q744. Which communication method best supports rapid team coordination during emergencies?
A. Written memos
B. Verbal briefings
C. Email summaries
D. Digital reports

Q745. Which training element ensures learners can demonstrate new skills effectively?
A. Practice opportunities
B. Long lectures
C. Reading lists
D. Written assignments

Q746. Which approach helps trainers identify training content gaps?
A. Peer review
B. Post-training survey
C. Needs analysis
D. Onboarding assessment

Q747. Which communication channel supports detailed, documented information sharing?
A. Formal reports
B. Face-to-face discussions
C. Verbal briefings
D. Instant messaging

Q748. Which principle improves long-term learning through review and reinforcement?
A. Overload
B. Spacing effect
C. Intensity
D. Isolation

Q749. Which type of communication is most effective for providing performance feedback?
A. One-on-one meetings
B. Group announcements
C. Email bulletins
D. Memos

Q750. Which concept emphasizes continuous learning and improvement in the workplace?

A. Static knowledge
B. Learning culture
C. Performance discipline
D. Regulatory compliance

Q751. What is the primary responsibility of a safety professional under legal compliance requirements?

A. Create new safety laws
B. Approve all employee disciplinary actions
C. Represent employees in legal disputes
D. Ensure all applicable regulations are identified and followed

Q752. Which document serves as the official record for workplace injuries and illnesses?

A. Injury and Illness Log
B. Incident Investigation Report
C. Risk Assessment Matrix
D. Hazard Identification Form

Q753. Why is it important for safety professionals to understand contract terminology?

A. To manage legal obligations and risk allocation effectively
B. To improve employee morale
C. To increase production efficiency
D. To enhance personal communication skills

Q754. Which type of law governs the relationship between private parties in safety matters?

A. Constitutional law
B. Criminal law
C. Administrative law
D. Civil law

Q755. What is the primary purpose of a written safety policy statement?

A. To communicate management's commitment to safety and compliance
B. To serve as a disciplinary tool
C. To satisfy insurance company requirements only
D. To reduce audit frequency

Q756. Which principle limits a safety professional's authority to their area of expertise?

A. Comparative negligence
B. Assumption of risk
C. Professional scope of practice
D. Duty of care

Q757. What legal concept holds employers responsible for the acts of their employees?

A. Vicarious liability
B. Comparative fault
C. Product liability
D. Contributory negligence

Q758. Why is accurate record retention critical in safety management?

A. To demonstrate compliance and provide evidence in legal cases
B. To reduce paperwork costs
C. To allow faster internal audits only
D. To improve teamwork

Q759. What is the primary legal purpose of worker compensation laws?

A. To fund company safety programs
B. To provide medical and wage benefits to employees injured on the job
C. To replace health insurance requirements
D. To penalize negligent workers

Q760. Which type of legal action involves violations of safety regulations enforced by government agencies?
A. Administrative law case
B. Civil lawsuit
C. Criminal prosecution
D. Contract dispute
Q761. What is the term for transferring risk through a written agreement with another party?
A. Contractual risk transfer
B. Shared liability
C. Implied assumption
D. Tort indemnity
Q762. Why must contractors be evaluated for safety performance before hiring?
A. To satisfy insurance auditors
B. To lower labor costs
C. To reduce liability for unsafe acts performed by contracted workers
D. To increase project speed
Q763. Which principle requires organizations to act reasonably to prevent harm to others?
A. Duty of care
B. Vicarious liability
C. Risk avoidance
D. Implied consent
Q764. Which type of insurance covers third-party injury or property damage caused by company operations?
A. General liability insurance
B. Worker compensation insurance
C. Professional indemnity insurance
D. Property insurance
Q765. What is the purpose of documenting safety training attendance?
A. To eliminate refresher training needs
B. To increase productivity scores
C. To satisfy voluntary reporting goals
D. To prove compliance and demonstrate employee competence
Q766. What is the legal term for failing to exercise reasonable care that results in harm?
A. Negligence
B. Liability
C. Tort
D. Breach of contract
Q767. Why is confidentiality important in occupational health records?
A. To protect employee privacy and maintain trust
B. To limit access to insurance providers
C. To ensure medical professionals share data
D. To simplify administrative filing
Q768. Which defense can an employer use when an employee voluntarily accepts known hazards?
A. Comparative negligence
B. Assumption of risk
C. Contributory fault
D. Dual capacity
Q769. What is the main function of a safety audit from a legal perspective?
A. To verify compliance and identify potential liabilities
B. To assess employee satisfaction

C. To measure productivity
D. To determine equipment cost
Q770. Why must employers report certain workplace incidents promptly?
A. To comply with regulatory notification requirements
B. To maintain public relations
C. To reduce insurance costs
D. To improve morale
Q771. Which type of evidence is most valuable in defending a safety professional during litigation?
A. Generic safety posters
B. Verbal statements by coworkers
C. Contemporaneous documentation of actions taken
D. Historical industry data
Q772. Which principle governs fair treatment and equal opportunity in workplace safety programs?
A. Non-discrimination
B. Contributory liability
C. Vicarious responsibility
D. Risk tolerance
Q773. What is the role of informed consent in occupational health testing?
A. To ensure employees voluntarily agree to testing after understanding procedures and risks
B. To reduce testing costs
C. To eliminate the need for documentation
D. To shorten evaluation time
Q774. Which legal principle requires safety professionals to act in the best interest of employers and employees?
A. Comparative liability
B. Fiduciary duty
C. Vicarious responsibility
D. Breach of contract
Q775. Why is contractor safety management crucial on multi-employer worksites?
A. To establish clear accountability and prevent cross-liability issues
B. To shorten project schedules
C. To reduce reporting paperwork
D. To increase profitability
Q776. Which principle requires employers to ensure a workplace free from recognized hazards?
A. General duty of care
B. Comparative fault
C. Assumed liability
D. Negligent entrustment
Q777. Why should safety professionals maintain detailed inspection records?
A. To justify budget requests
B. To support marketing initiatives
C. To replace training materials
D. To demonstrate compliance and document hazard correction actions
Q778. Which concept ensures that penalties correspond proportionally to the severity of a violation?
A. Fair enforcement principle
B. Duty of diligence
C. Comparative negligence
D. Vicarious accountability
Q779. Why is it essential to define safety responsibilities within contracts?
A. To clearly assign accountability and minimize ambiguity

B. To satisfy audit requirements only
C. To reduce cost negotiation time
D. To eliminate contractor competition

Q780. What is the key legal risk when failing to document employee safety training?
A. Excessive insurance premiums
B. Increased turnover rates
C. Inability to prove compliance in case of litigation
D. Reduced morale

Q781. What does the term "due diligence" signify for safety professionals?
A. Taking all reasonable steps to identify and control hazards
B. Delegating all safety tasks to subordinates
C. Waiting for regulatory inspection before acting
D. Minimizing documentation

Q782. Which document typically outlines roles, deliverables, and risk allocation between parties?
A. Job safety analysis
B. Contract agreement
C. Safety manual
D. Incident log

Q783. Why is confidentiality crucial during workplace incident investigations?
A. To protect sensitive data and prevent bias or retaliation
B. To conceal company information
C. To simplify documentation
D. To accelerate insurance claims

Q784. Which legal principle applies when both employer and employee share responsibility for an injury?
A. Absolute defense
B. Vicarious liability
C. Strict liability
D. Comparative negligence

Q785. Which element must be proven in a negligence claim?
A. Breach of duty leading to harm
B. Financial hardship
C. Contract violation
D. Intent to harm

Q786. What is the main function of risk transfer in safety contracts?
A. To assign liability for specific risks to another party
B. To eliminate all hazards
C. To create shared insurance pools
D. To reduce project documentation

Q787. Which type of insurance protects safety professionals from claims related to professional advice errors?
A. Property insurance
B. General liability insurance
C. Professional liability insurance
D. Worker compensation insurance

Q788. Why should safety professionals avoid dual relationships that create conflicts of interest?
A. To preserve objectivity and maintain ethical integrity
B. To increase profitability
C. To simplify task allocation
D. To reduce supervision needs

Q789. Which component of a safety program serves as a legal proof of hazard communication compliance?
A. Written hazard communication plan
B. Training attendance sheet
C. Equipment inventory
D. Audit checklist
Q790. What is the main objective of whistleblower protection laws?
A. To penalize supervisors for negligence
B. To protect employees who report safety violations in good faith
C. To eliminate audits
D. To reward compliance financially
Q791. Which principle ensures that legal actions are filed within a reasonable time frame?
A. Statute of limitations
B. Duty of loyalty
C. Implied warranty
D. Civil recourse
Q792. Why should companies maintain contractor orientation records?
A. To verify that contractors were informed of site-specific hazards
B. To support human resources reviews
C. To document employee performance
D. To satisfy payroll reporting
Q793. What is the main purpose of an ethics code for safety professionals?
A. To increase authority
B. To reduce workload
C. To standardize inspection tools
D. To guide conduct and ensure integrity in professional practice
Q794. Why must employee impairment policies be enforced uniformly?
A. To ensure fairness and reduce liability from inconsistent application
B. To reduce absenteeism rates
C. To improve public image
D. To streamline training
Q795. Which principle holds that ignorance of the law is not an acceptable defense?
A. Duty to know
B. Assumed compliance
C. Legal accountability principle
D. Vicarious obligation
Q796. What is the legal benefit of maintaining accurate exposure records?
A. To establish evidence for future occupational illness claims
B. To reduce health monitoring frequency
C. To replace medical evaluations
D. To shorten audit duration
Q797. Why must management review and sign off on safety policies annually?
A. To create new job titles
B. To reaffirm commitment and ensure policies reflect current regulations
C. To satisfy vendor requirements
D. To prepare for insurance renewals
Q798. Which liability arises from producing or selling defective safety equipment?
A. Product liability
B. Employer negligence
C. Comparative fault

D. Strict accountability

Q799. Why is it important to document corrective actions following safety audits?

A. To verify that identified issues were resolved and compliance maintained

B. To impress clients

C. To reduce training sessions

D. To justify resource allocation

Q800. What is the key purpose of legal compliance audits?

A. To analyze worker productivity

B. To assess equipment lifespan

C. To review marketing plans

D. To evaluate adherence to applicable laws and regulations

Answer keys and Explanations (701-800)

Q701. Answer: B

Explanation: Concise language promotes clarity by delivering information in a straightforward and focused manner. It minimizes misunderstanding and improves retention, especially when communicating safety or operational procedures. Clear and concise communication ensures all employees interpret messages uniformly, reducing potential errors and confusion.

Q702. Answer: B

Explanation: Scaffolding helps learners relate new concepts to prior experience, gradually reducing support as competence increases. This technique builds confidence and understanding by providing structure during the learning process. It encourages independent problem-solving and ensures skill mastery through guided progression.

Q703. Answer: A

Explanation: Simulation role-play provides real-life communication practice in a risk-free environment. It enhances interpersonal skills, empathy, and leadership behavior by mimicking workplace interactions. This experiential learning method improves confidence and reinforces effective communication strategies for supervisory roles.

Q704. Answer: B

Explanation: Observation allows trainers or supervisors to assess skill application and behavioral changes in real work settings. It measures whether training translates into effective job performance. Direct observation provides authentic evidence of learning outcomes and identifies areas needing reinforcement.

Q705. Answer: D

Explanation: Blended learning integrates online modules with classroom instruction to create flexibility and engagement. Learners benefit from self-paced study complemented by interactive discussions. This balance supports diverse learning styles and maximizes understanding and retention of course content.

Q706. Answer: B

Explanation: Open participation fosters dialogue and inclusivity, ensuring all team members contribute to discussions. It builds shared understanding and trust, promoting collaboration and ownership of safety practices. Encouraging input leads to better hazard identification and problem-solving across the workforce.

Q707. Answer: B

Explanation: Andragogy, developed by Malcolm Knowles, emphasizes that adults prefer self-directed learning driven by internal motivation. Adults value autonomy and practical application of knowledge to solve real-life problems. This theory guides training programs to focus on relevance, experience, and learner responsibility.

Q708. Answer: A

Explanation: Immediate corrective feedback reinforces correct behaviors while promptly addressing errors during skill practice. It enhances motor learning, accuracy, and confidence by providing real-time guidance. Quick feedback ensures learners adjust performance effectively, promoting faster skill mastery and retention.

Q709. Answer: B

Explanation: Learning objectives specify what learners will be able to do after training. They guide content development, instructional methods, and assessments. Clear objectives ensure alignment between training activities and organizational goals, providing benchmarks for evaluating success.

Q710. Answer: C

Explanation: Constructive praise acknowledges correct actions and reinforces desired behaviors. Positive reinforcement motivates employees to maintain high performance and adopt safe work practices. Recognition builds confidence and supports a culture of encouragement and continuous improvement.

Q711. Answer: B

Explanation: Microlearning delivers short, targeted modules that focus on specific skills or concepts. This method fits modern attention spans and allows quick, efficient learning. It improves retention by enabling learners to digest and apply information incrementally without cognitive overload.

Q712. Answer: B

Explanation: Relevance connects learning content to learners' real-world experiences and job roles. When learners understand how training applies directly to their work, they are more engaged and motivated. This principle improves long-term retention and facilitates practical application.

Q713. Answer: D

Explanation: Noise—both physical and psychological—distorts message transmission and understanding. Distractions, background sounds, or emotional stress can interfere with communication effectiveness. Minimizing noise ensures that critical safety and operational messages are accurately received and acted upon.

Q714. Answer: A

Explanation: E-learning allows remote teams to access consistent training content anytime and anywhere. It reduces logistical constraints and ensures standardization across locations. This flexible delivery model supports global organizations and enhances learner convenience and engagement.

Q715. Answer: A

Explanation: Self-efficacy refers to an individual's confidence in their capacity to execute specific actions effectively. High self-efficacy improves task persistence, resilience, and performance. Training programs that build confidence through practice and feedback enhance this belief, resulting in better outcomes.

Q716. Answer: B

Explanation: Open dialogue encourages transparent discussions about risks, solutions, and safety practices. It builds trust and empowers employees to share concerns without fear. This two-way communication reinforces a proactive safety culture centered on collaboration and mutual responsibility.

Q717. Answer: B

Explanation: Maintaining appropriate eye contact communicates attentiveness, confidence, and respect. It enhances message credibility and fosters connection between speaker and listener. Effective use of eye contact supports engagement and mutual understanding during communication exchanges.

Q718. Answer: C

Explanation: Return on investment (ROI) analysis quantifies the financial impact of training by comparing benefits gained to program costs. It reveals performance improvements through measurable data such as efficiency, error reduction, or safety incidents. ROI validates training's contribution to organizational goals.

Q719. Answer: B

Explanation: Interactive engagement tools like polls, chats, and breakout rooms keep learners actively involved online. They enhance focus and participation, preventing disengagement. Interactive features transform virtual training into dynamic, learner-centered experiences that promote retention and collaboration.

Q720. Answer: A

Explanation: Adults prefer task-focused learning that addresses immediate challenges or problem-solving needs. Training built around real scenarios increases relevance and engagement. This practical focus ensures learners can directly apply knowledge to job situations, enhancing efficiency and retention.

Q721. Answer: C

Explanation: Facilitators guide and support learners by creating opportunities for exploration and discovery. They encourage interaction, critical thinking, and collaboration rather than solely transmitting information. This learner-centered role empowers adults to take responsibility for their learning outcomes.

Q722. Answer: B

Explanation: Journaling prompts learners to reflect on their experiences and identify lessons learned. This introspective method deepens understanding and promotes personal growth. Reflection encourages awareness of strengths, challenges, and progress toward performance goals.

Q723. Answer: D

Explanation: Adapting language to fit cultural and linguistic differences ensures messages are understood by all audiences. Using simple wording, visuals, and translations enhances inclusivity. This approach fosters equal understanding and participation in safety communication efforts.

Q724. Answer: B

Explanation: Collaborative communication involves employees in decision-making, making them feel valued and responsible. It enhances commitment to safety policies and encourages teamwork. This participative approach strengthens motivation and reinforces shared accountability for safety performance.

Q725. Answer: A

Explanation: Personalized learning tailors content, pace, and delivery to individual learner preferences and abilities. It maximizes engagement and effectiveness by addressing specific skill gaps. Customization enhances satisfaction, accelerates learning, and improves overall performance outcomes.

Q726. Answer: B

Explanation: Interleaving involves mixing different types of learning activities or subjects rather than focusing on one topic continuously. This technique enhances long-term retention and skill transfer by challenging the brain to distinguish and connect diverse concepts. It improves problem-solving flexibility and reinforces deeper understanding.

Q727. Answer: A

Explanation: A feedback loop ensures two-way communication, confirming the message was received and interpreted correctly. It allows clarification, adjustment, and validation of understanding. This mechanism builds trust and enhances message accuracy throughout the organization.

Q728. Answer: C

Explanation: Social learning theory, proposed by Albert Bandura, highlights that adults learn by observing others and modeling behaviors. It shows that reinforcement and example-based instruction strongly influence skill adoption. This approach is widely used in safety and leadership training programs.

Q729. Answer: A

Explanation: Level 1 of Kirkpatrick's model evaluates participant reactions to training, including engagement, relevance, and satisfaction. Positive reactions indicate effective delivery and perceived value. Measuring this level helps trainers refine content and delivery methods for improved learner experience.

Q730. Answer: B

Explanation: Frequent and honest updates maintain transparency and trust during change. Open communication reduces uncertainty, increases employee involvement, and minimizes resistance. Consistent information sharing demonstrates leadership credibility and supports smoother transitions.

Q731. Answer: D

Explanation: Interactive workshops promote teamwork, discussion, and problem-solving among participants. They foster deeper understanding through shared experiences and group reflection. Collaboration encourages critical thinking and helps learners apply knowledge to realistic workplace situations.

Q732. Answer: B

Explanation: A Learning Management System (LMS) centralizes training content, completion records, and progress tracking. It provides analytics for evaluating learning outcomes and compliance. LMS tools enhance accountability, accessibility, and consistency in employee development programs.

Q733. Answer: B

Explanation: Customization tailors training materials to accommodate visual, auditory, and kinesthetic learning preferences. It ensures inclusivity and maximizes engagement by addressing diverse learner needs. This personalized approach improves understanding, retention, and training effectiveness.

Q734. Answer: B

Explanation: Phone calls enable instant dialogue and immediate feedback between communicators. This real-time exchange allows clarification and ensures message accuracy. Direct communication strengthens relationships and resolves issues efficiently compared to asynchronous methods.

Q735. Answer: D

Explanation: Alignment guarantees that learning activities, content, and assessments directly support training objectives. It creates coherence between instructional intent and learner outcomes. Proper alignment ensures focused training that delivers measurable performance improvements.

Q736. Answer: C

Explanation: Simplifying language ensures inclusivity and comprehension across diverse cultural backgrounds. Avoiding jargon and idiomatic expressions promotes clear communication and participation. This approach minimizes misinterpretation and fosters stronger collaboration in multicultural teams.

Q737. Answer: A

Explanation: Guided discovery allows learners to explore problems and find solutions with instructor support. It encourages critical thinking and deeper understanding through experiential engagement. This approach fosters independence and long-term retention by promoting active learning.

Q738. Answer: B

Explanation: Active listening requires focusing fully on the speaker, understanding their message, and providing thoughtful feedback. It builds trust, reduces misunderstandings, and strengthens interpersonal relationships. Practicing active listening enhances workplace collaboration and communication quality.

Q739. Answer: A

Explanation: Collaboration fosters knowledge sharing, mutual respect, and efficiency across departments. It ensures that training content addresses diverse operational perspectives. Cross-functional cooperation enhances organizational learning and overall performance.

Q740. Answer: A

Explanation: E-learning enables learners to access content at their convenience, accommodating individual schedules and learning speeds. It promotes consistency and accessibility while allowing learners to review material as needed. This flexibility enhances engagement and retention.

Q741. Answer: D

Explanation: Nodding communicates attentiveness, agreement, or understanding during dialogue. It reinforces connection between speaker and listener, promoting positive communication flow. Nonverbal cues like nodding enhance engagement and encourage open participation.

Q742. Answer: B

Explanation: Soft skills training develops interpersonal, communication, and emotional intelligence capabilities. It equips leaders to influence, motivate, and collaborate effectively. These skills complement technical expertise and strengthen organizational culture and productivity.

Q743. Answer: B

Explanation: Autonomy empowers adult learners to direct their learning path, pace, and methods. Providing choice enhances motivation and ownership. This independence encourages deeper engagement and fosters continuous professional development.

Q744. Answer: B

Explanation: Verbal briefings allow instant information exchange and clarification in high-pressure situations. They ensure swift coordination and alignment of team actions. Immediate verbal updates are critical for accuracy and responsiveness during emergencies.

Q745. Answer: A

Explanation: Practice opportunities give learners hands-on experience to apply and refine new skills. Repetition and feedback reinforce correct performance and confidence. This experiential approach bridges theory and practice, ensuring skill proficiency.

Q746. Answer: C

Explanation: Needs analysis determines the gap between current and desired performance levels. It identifies specific knowledge and skill areas requiring development. This process ensures training is targeted, efficient, and aligned with organizational objectives.

Q747. Answer: A

Explanation: Formal reports provide structured, documented communication that captures details accurately. They serve as records for reference, compliance, and accountability. Written documentation ensures consistency and traceability of complex information.

Q748. Answer: B

Explanation: The spacing effect enhances memory retention by reviewing information over intervals rather than all at once. Periodic reinforcement solidifies understanding and reduces forgetting. This principle is vital for continuous learning and performance maintenance.

Q749. Answer: A

Explanation: One-on-one meetings create a confidential, supportive setting for personalized feedback. They allow discussion, clarification, and goal-setting between manager and employee. This approach strengthens trust and promotes targeted development.

Q750. Answer: B

Explanation: A learning culture values ongoing education, skill enhancement, and shared knowledge. It encourages curiosity, innovation, and adaptability across all levels of an organization. Establishing a learning culture ensures sustained growth and operational excellence.

Q751. Answer: D

Explanation: The safety professional's primary legal responsibility is to identify, interpret, and implement applicable safety and health regulations. This includes ensuring compliance with federal, state, and local standards. By proactively monitoring changes in legislation and maintaining compliance documentation, the professional prevents legal violations and helps protect the organization from potential penalties and litigation.

Q752. Answer: A

Explanation: The Injury and Illness Log serves as the official record of occupational injuries and illnesses within a workplace. It provides a standardized means to record the type, date, and severity of each case. Maintaining accurate logs demonstrates compliance with recordkeeping requirements and helps identify trends for preventive measures.

Q753. Answer: A

Explanation: Understanding contract terminology enables safety professionals to identify responsibilities, obligations, and liability clauses that affect safety performance. Contracts often outline who is responsible for compliance, risk transfer, and insurance coverage. Proper comprehension ensures safety requirements are clearly defined and helps prevent disputes related to workplace incidents.

Q754. Answer: D

Explanation: Civil law governs disputes between private parties, including those involving negligence or breach of contract in safety management. It defines liability when one party's actions or omissions cause harm to another. Safety professionals must understand civil liability to minimize exposure through proper documentation, training, and compliance practices.

Q755. Answer: A

Explanation: A written safety policy communicates the organization's leadership commitment to maintaining a safe workplace and complying with legal standards. It outlines responsibilities, establishes accountability, and provides a framework for safety goals. A clear policy demonstrates due diligence and supports defense against negligence claims.

Q756. Answer: C

Explanation: A professional scope of practice defines the boundaries within which a safety professional is qualified to act. Staying within this scope ensures decisions are based on competent knowledge and prevents liability for actions beyond one's expertise. It reinforces ethical conduct and professional credibility in legal contexts.

Q757. Answer: A

Explanation: Vicarious liability makes an employer legally responsible for negligent acts committed by employees within the scope of their employment. This principle ensures organizations maintain proper supervision, training, and safety systems to prevent harm, reinforcing accountability at the management level.

Q758. Answer: A

Explanation: Accurate record retention ensures documentation of inspections, training, and incident reports is available for regulatory review or litigation. Properly maintained records demonstrate due diligence and can serve as crucial evidence in defending the organization's actions following a workplace incident or compliance audit.

Q759. Answer: B

Explanation: Worker compensation laws provide financial and medical support to employees who suffer work-related injuries or illnesses. These laws protect both employees and employers by ensuring prompt benefits while limiting lawsuits. Compliance ensures fair treatment and supports an effective recovery process for injured workers.

Q760. Answer: A

Explanation: Administrative law cases arise when regulatory agencies enforce compliance with established safety standards. These proceedings can result in citations, penalties, or corrective orders. Safety professionals must understand administrative procedures to manage compliance and effectively respond to enforcement actions.

Q761. Answer: A

Explanation: Contractual risk transfer involves assigning specific responsibilities and liabilities to another party through written agreements. This practice ensures that contractors or vendors bear appropriate responsibility for safety performance, reducing organizational exposure to potential legal and financial risks associated with workplace incidents.

Q762. Answer: C

Explanation: Evaluating contractor safety performance before hiring ensures that only qualified and compliant contractors perform work on-site. This reduces the organization's liability for unsafe acts, protects all parties, and maintains consistent safety standards across multi-employer worksites.

Q763. Answer: A

Explanation: Duty of care is the legal obligation requiring organizations to act with reasonable caution to prevent harm. It applies to management decisions, workplace design, and employee supervision. Breaching this duty can lead to negligence claims, emphasizing the need for proactive safety management and ethical conduct.

Q764. Answer: A

Explanation: General liability insurance protects a company against claims involving third-party injuries, property damage, or personal harm caused by business operations. It supports financial stability and risk management by covering legal defense and settlement costs related to such claims.

Q765. Answer: D

Explanation: Documenting safety training attendance verifies that employees received required instruction on workplace hazards and procedures. These records prove regulatory compliance, confirm worker competence, and provide legal protection by demonstrating that the employer took reasonable steps to prevent accidents.

Q766. Answer: A

Explanation: Negligence occurs when an individual or organization fails to act with the level of care expected under similar circumstances, causing harm. It forms the basis for many civil lawsuits. Preventing negligence requires adherence to established standards, proper supervision, and documentation of preventive measures.

Q767. Answer: A

Explanation: Confidentiality protects employees' personal and medical information, fostering trust and compliance with privacy laws. Restricting access ensures sensitive data is handled responsibly and only by authorized personnel. This promotes ethical management of workplace health information and reduces liability risks.

Q768. Answer: B

Explanation: Assumption of risk applies when an employee knowingly engages in work despite being aware of inherent hazards. While rarely a complete defense today, it emphasizes the importance of hazard communication and informed consent in preventing claims of employer negligence.

Q769. Answer: A

Explanation: A safety audit evaluates compliance with legal and organizational standards, uncovering potential areas of liability. It ensures safety programs are functioning effectively and documents proactive management efforts. Regular audits demonstrate due diligence and reduce exposure to citations or lawsuits.

Q770. Answer: A

Explanation: Prompt reporting of workplace incidents is a legal requirement to ensure timely investigation and corrective action. Regulatory agencies rely on accurate reporting to monitor compliance and workplace safety trends. Failure to report can result in significant fines and reputational damage.

Q771. Answer: C

Explanation: Contemporaneous documentation—records created at the time of an event—provides strong evidence that appropriate actions were taken. It supports credibility and demonstrates consistent, proactive safety management. Accurate documentation is often decisive in defending against negligence or liability claims.

Q772. Answer: A

Explanation: The principle of non-discrimination ensures that safety programs apply equally to all employees, regardless of background or status. It promotes fairness and inclusivity, fostering a culture of respect while ensuring compliance with equal employment and human rights legislation.

Q773. Answer: A

Explanation: Informed consent ensures employees voluntarily agree to health testing after understanding the purpose, process, and potential outcomes. It safeguards personal rights, builds transparency, and ensures compliance with ethical and legal standards in workplace health assessments.

Q774. Answer: B

Explanation: Fiduciary duty refers to the ethical and legal responsibility to act loyally and in good faith for the benefit of those served, such as employers and employees. Safety professionals uphold this duty by ensuring honesty, integrity, and diligence in their professional conduct.

Q775. Answer: A

Explanation: Contractor safety management ensures all parties understand their safety roles, responsibilities, and communication protocols. It prevents incidents arising from coordination failures and limits cross-liability between employers. Proper oversight fosters accountability and promotes a unified safety culture across multi-employer worksites.

Q776. Answer: A

Explanation: The general duty of care requires employers to provide a workplace free from recognized hazards likely to cause serious harm. It mandates reasonable preventive action and risk control measures. This principle emphasizes proactive safety management, ensuring employers anticipate potential dangers before they result in employee injury or illness.

Q777. Answer: D

Explanation: Detailed inspection records provide evidence that safety inspections were conducted and corrective actions taken. These documents are essential for demonstrating compliance during audits or legal proceedings. They prove management diligence and serve as vital tools for tracking hazard trends and preventing recurrence.

Q778. Answer: A

Explanation: The fair enforcement principle requires that penalties for safety violations reflect the seriousness of the infraction and the employer's level of negligence. This proportional approach promotes justice while motivating continuous compliance and improvement. It ensures enforcement actions remain corrective rather than punitive in intent.

Q779. Answer: A

Explanation: Defining safety responsibilities in contracts ensures that all parties understand their obligations regarding compliance, hazard control, and worker protection. Clear allocation of duties prevents confusion, reduces legal disputes, and strengthens the organization's ability to enforce safety standards across project partners.

Q780. Answer: C

Explanation: Failure to document safety training eliminates proof that employees were informed about hazards and safe practices. In litigation or enforcement actions, lack of documentation can be interpreted as negligence. Proper records provide evidence of compliance, training effectiveness, and due diligence in protecting workers.

Q781. Answer: A

Explanation: Due diligence means taking every reasonable measure to anticipate, recognize, and mitigate workplace hazards. It demonstrates proactive management and serves as a legal defense by showing the employer acted responsibly to protect employees. Consistent diligence reduces risks and strengthens legal compliance.

Q782. Answer: B

Explanation: A contract agreement formalizes responsibilities, deliverables, and risk-sharing mechanisms between parties. It defines who is accountable for safety compliance, performance standards, and liability coverage. Properly structured contracts reduce conflicts and ensure that safety requirements are legally binding.

Q783. Answer: A

Explanation: Maintaining confidentiality during incident investigations safeguards employee privacy and encourages honest participation. It prevents retaliation, preserves evidence integrity, and builds trust between management and workers. Ethical handling of information ensures fairness and strengthens the investigation's credibility.

Q784. Answer: D

Explanation: Comparative negligence assigns liability proportionally based on each party's contribution to an incident. This principle recognizes that both the employer's and employee's actions may influence the outcome. It promotes balanced accountability and encourages both parties to exercise caution and responsibility.

Q785. Answer: A

Explanation: To establish negligence, the claimant must prove the defendant owed a duty of care, breached that duty, and caused actual harm. The link between breach and injury is crucial. Demonstrating these elements substantiates liability and helps determine compensation for damages.

Q786. Answer: A

Explanation: Risk transfer in contracts allocates responsibility for particular risks to another entity, such as a contractor or insurer. It clarifies who bears financial or legal consequences if an incident occurs. Properly executed risk transfer reduces exposure and strengthens organizational resilience against potential claims.

Q787. Answer: C

Explanation: Professional liability insurance covers safety professionals against claims arising from errors, omissions, or negligence in their professional duties. It provides financial protection and supports legal defense, ensuring the professional can continue practicing responsibly while maintaining trust and credibility.

Q788. Answer: A

Explanation: Avoiding dual relationships prevents bias and ensures that safety professionals act solely in the interest of safety and compliance. Conflicts of interest can compromise judgment, damage credibility, and lead to ethical violations. Maintaining neutrality supports fairness and professional standards.

Q789. Answer: A

Explanation: A written hazard communication plan documents how chemical hazards are identified, labeled, and communicated to workers. It demonstrates compliance with legal requirements and serves as evidence during inspections. The plan ensures all employees receive consistent, accurate safety information.

Q790. Answer: B

Explanation: Whistleblower protection laws shield employees who report safety violations or unethical practices from retaliation. They promote transparency and accountability by encouraging employees to speak up about unsafe or unlawful conditions without fear of adverse consequences.

Q791. Answer: A

Explanation: The statute of limitations defines the time limit within which a legal claim must be filed. This principle ensures fairness by requiring timely action while evidence and witness memories remain reliable. It protects both claimants and defendants from indefinite legal uncertainty.

Q792. Answer: A

Explanation: Contractor orientation records confirm that external workers received safety briefings about site-specific risks and procedures. These documents demonstrate compliance, accountability, and diligence in managing third-party personnel under the company's operational control.

Q793. Answer: D

Explanation: A code of ethics establishes standards of honesty, integrity, and professionalism that govern safety practice. It provides guidance in decision-making, promotes public trust, and ensures consistency in ethical behavior, reinforcing the credibility of the profession.

Q794. Answer: A

Explanation: Uniform enforcement of impairment policies guarantees equal treatment for all employees, avoiding discrimination claims and maintaining credibility. Consistency ensures workplace safety by addressing risks from drugs, alcohol, or fatigue objectively and lawfully.

Q795. Answer: C

Explanation: The legal accountability principle asserts that individuals and organizations are responsible for knowing and complying with applicable laws. Claiming ignorance does not exempt one from liability. Continuous education and awareness are essential to maintain compliance and prevent violations.

Q796. Answer: A

Explanation: Accurate exposure records serve as critical evidence if occupational illness claims arise years later. They provide a traceable history of potential exposures and demonstrate that monitoring and controls were maintained. This documentation supports legal defense and ensures accountability in workplace health management.

Q797. Answer: B

Explanation: Annual management review and approval of safety policies confirm leadership commitment to compliance and continuous improvement. Regular updates ensure policies align with new legal standards, industry best practices, and organizational goals, sustaining effective governance and accountability.

Q798. Answer: A

Explanation: Product liability holds manufacturers or sellers legally responsible for harm caused by defective or unsafe products. This ensures safety equipment meets performance standards and motivates producers to maintain rigorous quality control and testing procedures.

Q799. Answer: A

Explanation: Documenting corrective actions demonstrates that audit findings were addressed promptly, ensuring sustained compliance and system improvement. It provides evidence of accountability, supports due diligence, and reduces potential liability by showing proactive hazard management.

Q800. Answer: D

Explanation: Legal compliance audits systematically review an organization's adherence to applicable safety laws, regulations, and internal standards. These audits identify gaps, verify documentation accuracy, and drive corrective measures to maintain continuous compliance and prevent legal exposure.

Audio Book

Flash-Cards

Made in the USA
Las Vegas, NV
16 July 2026

51122629R00142